Water

Brian Kennedy, *Book Editor*

GREENHAVEN PRESS
A part of Gale, Cengage Learning

Detroit • New York • San Francisco • New Haven, Conn • Waterville, Maine • London

Elizabeth Des Chenes, *Managing Editor*

© 2012 Greenhaven Press, a part of Gale, Cengage Learning

Gale and Greenhaven Press are registered trademarks used herein under license.

For more information, contact:
Greenhaven Press
27500 Drake Rd.
Farmington Hills, MI 48331-3535
Or you can visit our Internet site at gale.cengage.com

ALL RIGHTS RESERVED.
No part of this work covered by the copyright herein may be reproduced, transmitted, stored, or used in any form or by any means graphic, electronic, or mechanical, including but not limited to photocopying, recording, scanning, digitizing, taping, Web distribution, information networks, or information storage and retrieval systems, except as permitted under Section 107 or 108 of the 1976 United States Copyright Act, without the prior written permission of the publisher.

For product information and technology assistance, contact us at

Gale Customer Support, 1-800-877-4253
For permission to use material from this text or product, submit all requests online at www.cengage.com/permissions

Further permissions questions can be e-mailed to permissionrequest@cengage.com

Articles in Greenhaven Press anthologies are often edited for length to meet page requirements. In addition, original titles of these works are changed to clearly present the main thesis and to explicitly indicate the author's opinion. Every effort is made to ensure that Greenhaven Press accurately reflects the original intent of the authors. Every effort has been made to trace the owners of copyrighted material.

Cover image Stockbyte/Getty Images.

LIBRARY OF CONGRESS CATALOGING-IN-PUBLICATION DATA

Water / Brian Kennedy, book editor.
   p. cm. -- (Issues that concern you)
Includes bibliographical references and index.
ISBN 978-0-7377-5702-6 (hardcover)
1. Water--Enviromental aspects. 2. Water use. 3. Water supply. 4. Water--Pollution. I. Kennedy, Brian.
GB661.2.W364 2011
363.6'1--dc23
                                                      2011020634

Printed in the United States of America
1 2 3 4 5 6 7 15 14 13 12 11

CONTENTS

Introduction   5

1. Many Sources of Pollution Endanger the Oceans   10
   Mother Jones

2. The Lack of Clean Water Is a Global Crisis   15
   Paul Alois

3. Access to Water Is a Human Right   23
   Food and Water Watch

4. Making Water a Human Right Will Not Improve Access   28
   Elizabeth Dickinson

5. There Is a Water Crisis in the Middle East   31
   John Bohannon

6. Africa Suffers from a Water Crisis   39
   Tina Rosenberg

7. Scientific Advances Take the Salt out of Salt Water   47
   Robert F. Service

8. The Lack of Adequate Sewage Systems Causes Major Health Problems   54
   Barbara Frost

9. Recycling Sewage Water Is Clean and Effective   60
   Eilene Zimmerman

10. Bottled Water Is a Huge Industry That Continues to Grow   65
    IBISWorld

11.	Drinking Bottled Water May Hurt the Environment *Tom Paulson*	72
12.	The Demand for Ethanol Is Creating a Water Problem *Jim Moscou*	79

Appendix

What You Should Know About Water	85
What You Should Do About Water	88

Organizations to Contact   92

Bibliography   97

Index   100

Picture Credits   104

INTRODUCTION

People need water. Not only does everybody need to drink it to survive, but many daily activities rely on water. People use it to wash clothes and dishes and to water their yards. Communities also depend on water in ways that are not as obvious. Water is necessary for sewage and waste removal; it allows for hygienic use of toilets instead of outhouses. All societies rely on water, but what would happen if supplies dried up?

California and the western United States have experienced such a crisis in recent years. There is not enough water for everyone. Population growth has increased the strain on local water supplies as more and more users tap into the same sources. Conservationists have advocated for new housing developments to account for their effects on the water supply, namely how much they will need and where it will come from. Environmental groups have successfully sued to block new housing developments that did not properly evaluate long-term impacts on water in the region. Tim Barnett, a professor at the University of California, San Diego (UCSD), warns against excessive growth: "If you want to continue to grow, where are you going to get the water? There has to be limits."[1]

This antidevelopment trend has its opponents. They claim that building is necessary to support jobs and sustain the economy. Joel Kotkin of Chapman University argues that if "California wants to become a place that can't accommodate people, people will continue to leave."[2]

Making matters worse, experts say global warming has reduced the annual snowmelt in the Rocky Mountains and the Sierra Nevada. California water officials expect the snowpack in the Sierra Nevada to decrease 25 to 40 percent by 2050. This in turn affects lakes and reservoirs. Lake Mead, a huge reservoir outside of Las Vegas, Nevada, could be dry within ten years according to estimates from researchers at UCSD. The Colorado River, which is fed by runoff from the Rockies, is a major water supply for seven western states. As runoff decreases and reduces the amount of water available, those states will have to address water rights.

Further complicating the water battle are environmental issues. Courts in California have restricted access to the water in the San Joaquin Delta in Northern California to protect endangered fish that live in the area. Environmentalists praise the ruling, while many Southern California residents and central California farmers complain that the needs of people are more important than those of the fish. Without sufficient water, farmers cannot utilize all of their land, causing losses of both profits and jobs. The California Farm Bureau reports that in 2008 farmers abandoned over one hundred thousand acres. Richard Howitt, a professor at the University of California, Davis (UC Davis) in the Central Valley, estimates that these reductions will eliminate thirty thousand to sixty thousand jobs and $1.6 billion in profits.

In addition to experiencing population growth, climate change, and court-ordered reductions in water withdrawals, California is in the midst of a drought. Conditions are dire, and something must be done to address the water shortage.

An obvious way to combat a water shortage is to look for other sources of water. Las Vegas—the fastest-growing major city in the United States—developed a plan to create a pipeline to deliver water to the city from rural northern Nevada. Ranchers in this area protested, arguing that their groundwater will dry up quickly if it is pumped to provide water for the city. Without water, their livelihood would be threatened. Another such pipeline was proposed by Utah, to deliver water to its southwestern corner. This water would be piped in from Nevada's Lake Powell, but people in Nevada oppose the idea.

Another method to generate more potable water is desalination, the process of removing the salt from seawater. Desalination allows coastal areas to access the vast amounts of readily available water in the oceans. The technology is already used in many countries around the world, and it is becoming more efficient and affordable as it improves. A desalination facility was scheduled to be completed in Carlsbad, California—just north of San Diego—in 2012. An older facility in Yuma, Arizona, that filters salty irrigation

California courts have restricted access to the water in the San Joaquin delta to protect endangered fish in the area. Environmental activists praise the ruling, but farmers condemn it.

runoff was reactivated in 2010 for a one-year test run. The trial was deemed a success, and the plant is likely to remain operational.

Opponents of desalination argue that the price is too high. Treated seawater costs many times more than regular tap water, not only because the technology is expensive to create and install, but also because the plants use large amounts of energy to operate. Critics also cite environmental concerns. Desalination facilities can harm sea life and the surrounding coastal environments, both by destroying marine life when taking in ocean water and by essentially poisoning it by discharging vast quantities of excess salt back into local waters. The Carlsbad facility was delayed for more than a decade by environmental challenges.

Pipelines and desalination are examples of ways to increase the supply of water. The other side of the equation is reducing the demand. The less water people use, the less strain they put on the local water supplies.

Some areas have forced residents to use less water by instituting water rationing. In 2009 San Diego mandated water rationing to reduce consumption by 8 percent. The city set an outdoor watering schedule and threatened fines against water wasters. People would be notified of problems and ordered to fix them. However, repeat offenders could be fined over one thousand dollars. Mike Bresnahan, a city water official, supports the plan: "We will insist that (water wasters) start making progress. If there is not compliance, we will continue to ratchet up the penalties."[3]

Los Angeles, California, enforces a similar program, requiring that outdoor sprinkler systems only be used three times per week, on specified days, and never between 9 A.M. and 5 P.M. Department of Water and Power officers patrol the city, issuing citations to violators. Ironically, the shift in water pressure on Mondays and Thursdays (citywide watering days) eventually caused many of the city's aging water mains to break.

Water rationing has its critics. They are quick to point out that enforcement is inconsistent; there are not nearly enough officers to catch all the violators. They argue that rate structures are more effective: charging people significantly more for water usage above certain thresholds encourages them to conserve. Bob Cook, general manager at the Lakeside, California, Water District, agrees. "I don't like the approach that you are just going to rely on the water-use restrictions," he says. "I just don't think it can be enforced that effectively."[4]

Water usage can be reduced in other ways. The Pacific Institute published a report in 2010 detailing the vast savings achievable by improving water-use efficiency. Residents should replace old water-using devices with newer, high-efficiency models and replace water-consuming plants with those that need minimal water. Farmers can install modern, more efficient irrigation systems. These changes will cost money up front but will result in massive reductions in water use—and water bills. The institute argues that conservation and efficiency should be central to any solutions to California's water problems.

California's government is trying to take action. A measure is on the 2012 ballot that would allocate $11 billion to overhaul the state's water system. The measure was slated to be on the 2010 ballot but was delayed due to the state's severe budget problems. Supporters claim that the bill allocates the resources necessary to address the water crisis. Opponents of the bill argue that it is too expensive and that much of the money will go to specific projects that only benefit a small number of people. Environmental groups like the Sierra Club also oppose it because it does not focus enough on conservation.

The water crisis in California and the West presents many problems. There are a variety of potential solutions, although with all the competing interests involved, none is simple. Many different people with many different interests have stakes in the outcome. They do have one concern in common: They all need water.

This anthology considers a variety of water issues and their impacts both on individuals and society. Additionally, the magazine articles, news reports, and excerpts from other sources provide varying perspectives. This volume also contains several appendixes to guide the reader to further research the topic. A thorough bibliography lists both books and articles for additional reading, and a list of organizations provides contacts for extra information. The appendix "What You Should Know About Water" lists important facts and statistics for quick reference, and "What You Should Do About Water" offers advice on understanding the issues and advises readers on what they can do to help alleviate the problems. *Issues That Concern You: Water* is an excellent resource and a great starting point for anyone interested in this important topic.

Notes

1. Douglas Quan, Kimberly Pierceall, and Janet Zimmerman. "Crisis on Tap: California's Water Reckoning." *The Press Enterprise*, March 22, 2009. www.pe.com/reports/2009/water/stories/PE_News_Local_S_water22.19771d4.html
2. Douglas Quan, Kimberly Pierceall, and Janet Zimmerman. "Crisis on Tap: California's Water Reckoning." *The Press Enterprise*, March 22, 2009.
3. Mike Lee. "Water Rationing Made Mandatory," *San Diego Union Tribune*, April 29. 2009. www.signonsandiego.com/news/2009/apr/24/1m24water235154-water-rationing-made-mandatory/?uniontrib
4. Mike Lee. "Water Rationing Made Mandatory," *San Diego Union Tribune*, April 29. 2009. www.signonsandiego.com/news/2009/apr/24/1m24water235154-water-rationing-made-mandatory/?uniontrib

ONE

Many Sources of Pollution Endanger the Oceans

Mother Jones

Mother Jones is a nonprofit news organization that specializes in investigative, political, and social justice reporting. It is a bimonthly national magazine that is linked to a website featuring new, original reporting. In this question-and-answer article, *Mother Jones* explains the sources and effects of ocean pollution. According to the author, antipollution measures have helped the problem, but more must be done, particularly to deal with pollution sources that are far inland from the coastal waters they damage.

There are many [sources of water pollution], especially toxins that come from industrial and municipal wastewaters, runoff from farms and urban areas, and the erosion of soils. These toxins can include naturally-occurring chemicals that are present in higher concentrations because of human activities as well as new, man-made compounds such as DDT [a pesticide].

Other pollutants include biostimulants from sewage and industrial wastes; oil from runoffs, accidental spills, and oil and gas production; sediments from erosion caused by farming, forestry, mining, and development; plastics and other debris from ships, fishing nets, and containers; thermal pollution from the cool-

Mother Jones, "Marine Pollution: How the Ocean Became a Toxic Waste Dump," March 1, 2006. Copyright © 2006, *Mother Jones*. Reproduced by permission.

ing water that comes from power and industrial plants; human pathogens from sewage, urban runoff, and livestock; and finally, alien species that are introduced into a habitat by ships.

Q: That's a lot of pollutants. What are the most important?

A: Since the Clean Water Act was passed and reauthorized in the 1970s and 1980s, the most harmful pollutants have actually come from diffuse sources rather than direct discharges. For example, oil pollution from ships, accidental spills and production activities has been curtailed after a concerted effort by environmentalists and policymakers, but diffuse pollution from various land-based activities—for instance, urban runoff—has not.

Lake Como in Fort Worth, Texas, is just one of forty lakes and reservoirs in the area contaminated by urban runoff. The runoff usually consists of residue from black sealants sprayed on parking lots, driveways, and playgrounds.

Of these pollutants, nowadays coastal areas are most endangered by the introduction of excess nutrients that overwhelm the local ecosystem.

How does pollution enter the water from these diffuse sources?

Pollutants from diffuse sources include those released into the atmosphere by fossil-fuel and waste combustion, along with pesticides, toxic-waste products, nutrients, and sediments that enter the water as runoff from the land. According to a Pew report on water pollution, the latter is the primary source of pollution in coastal waters, which is, in turn, where the "most demonstrable effects on living resources occur."

What effects does water pollution have on life in the water?

Toxins, such as those from industrial wastewaters, can poison living organisms—causing disease and reproductive failure, and can also pose human health risks. On the other hand, organic wastes—such as nitrogen and phosphorous—can overload coastal

"Dear Diary - Today my spirits are uplifted. I finally see signs of civilization," cartoon by David Brown. www.CartoonStock.com.

habitats and cause serious depletion of dissolved oxygen supplies needed by marine animals. It's true that normally, these habitats need nutrients, but too much can over-stimulate the environment, creating more organic matter than the ecosystem can handle. These wastes can also stimulate algal blooms, which can often kill off other organisms in the area.

Meanwhile, sediments from land runoff or dredging can decrease the clarity of the water and smother and cause the loss of sea grasses and coral reefs, which can in turn alter the food chains that support fisheries in the area.

What measures have been taken so far?

As mentioned before, maritime pollution has subsided somewhat over the past 30 years, thanks to the Federal Water Pollution Control Act, which was passed in 1972, and reauthorized twice in the 1980s as the Clean Water Act. The law imposed uniform minimum federal standards for municipal and industrial wastewater treatment, and put limits on pollutants in discharges from industrial facilities—requiring plants to adopt up-to-date pollution-control technology.

The Clean Water Act also increased standards for waste treatment plants—under law, about $125 billion was spent between 1972 and 1992 creating new public treatment works. The ban of pesticides and other harmful chemicals such as DDT, PCBs, and lead additives in gasoline, have also helped to control maritime pollution.

Have these measures worked?

Yes, despite the fact that there has been a steady increase in population in the United States, as well as an increase in wastewater discharge, maritime pollution has decreased dramatically. Oxygen levels in New York Harbor, for instance, are now 50 percent higher than they were 30 years ago. In the Southern California Bight, off Los Angeles and San Diego, inputs of many pollutants have been reduced 90 percent or more over a 25-year period, and the ecosystem there—including kelp, fish, and seabird populations—has greatly recovered.

What are the next steps?

As noted above, the most serious water-quality problems come from diffuse sources, and those are more difficult to deal with. According to a Pew report on water pollution, slowing or reversing coastal pollution will, in particular, require management strategies for a variety of watersheds that are often far inland from coastal environments.

Some measures to reduce pollution could include: advanced treatment of municipal wastewaters; the reduction of nitrogen oxide emissions from power plants and vehicles; the control of ammonia emissions from animal feedlots; the more efficient use of fertilizers and manure; and the restoration of wetlands and floodplains that often act as nutrient traps for runoff.

The Lack of Clean Water Is a Global Crisis

Paul Alois

> The Arlington Institute is a nonprofit research organization that specializes in planning positive global futures and trying to influence rapid change. It advocates that effective planning for the future is enhanced by applying newly emerging technology. Paul Alois worked as a research analyst at the Arlington Institute. In the following viewpoint Alois explains that population growth, as well as numerous other contributing factors, has led to water shortages in many parts of the world. He contends that the problem will worsen as the world's population increases unless a concerted effort is taken to address and mitigate the issue.

Water, simply put, makes the existence of the human race on this planet possible. With few exceptions, water has always been a natural resource that people take for granted. Today, the situation has changed.

The World Bank [a global financial institution that provides loans to developing countries] reports that 80 countries now have water shortages and 2 billion people lack access to clean water. More disturbingly, the World Health Organization has reported that 1 billion people lack enough water to simply meet their basic needs.

Paul Alois, "Global Water Crisis Overview," The Arlington Institute, April 2007. Copyright © 2007, The Arlington Institute. Reproduced by permission.

Population growth and groundwater depletion present the two most significant dangers to global water stability. In the last century, the human population has increased from 1.7 billion people to 6.6 billion people, while the total amount of potable water has slightly decreased. Much of the population growth and economic development experienced in the last fifty years has been supported by subterranean water reserves called groundwater. These non-renewable reserves, an absolutely essential aspect of the modern world, are being consumed at an unsustainable rate.

The Present Supply and Usage of Water

Humanity has approximately 11 trillion cubic meters [m^3] of freshwater at its disposal. Groundwater aquifers contain over 95% of this water, while rain, rivers, and lakes make up the remaining 5%. Approximately 1,700 m^3 of water exists for every person on the planet, an alarming low number. According to the Water Stress Index, a region with less than 1,700 m^3 per capita is considered "water stressed."

The global supply is not distributed evenly around the planet, nor is water equally available at all times throughout the year. Many areas of the world have seriously inadequate access to water, and many places with high annual averages experience alternating seasons of drought and monsoons.

Water usage differs highly between developing countries and developed ones. Developing countries use 90% of their water for agriculture, 5% for industry, and 5% for urban areas. Developed countries use 45% of their water for agriculture, 45% for industry, and 10% for urban areas.

In the last century water usage per person doubled, even as the total population tripled, creating a situation today where many areas of the world are consuming water at an unsustainable rate.

Increasing Demand

The agricultural sector, by far the largest consumer of freshwater resources, accounts for 70% [of] global consumption. Irrigation consumes most of the water in the agricultural sector, and has

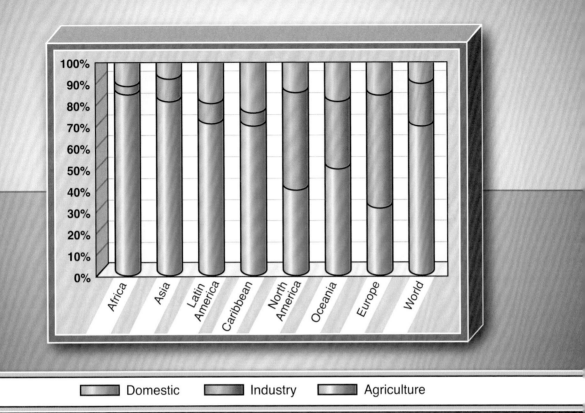

become an integral part of modern civilization because of access to groundwater aquifers. Once farmers were freed from relying on rain to water their crops, highly efficient commercial farming became increasingly common. This innovation also underpinned the Green Revolution [a series of initiatives undertaken between 1943 and the late 1970s], which dramatically increased crop production throughout the third world in the 1960s. Unfortunately, water is being drawn from many of these aquifers faster than it is being replaced.

The industrial sector accounts for 22% of global water consumption; this number will grow in the coming decades as the developing world industrializes. The needs of industry tend to take precedence over agriculture for simple economic reasons. 1,000 tons of water will produce 1 ton of wheat, which is worth $200. 1,000 tons of water in the industrial sector, however, will generate $14,000 worth of goods. On a per ton basis, industry creates 70 times more wealth. Despite its economic benefits, intense water use by industry has led to serious pollution that is beginning to create problems worldwide.

The residential sector uses the remaining 8% of the total water supply. Although this sector only accounts for a small percentage of overall use, it always takes precedence over industry and agriculture. In the last fifty years the world's urban population has exploded, and by 2010 50% of the people on the planet will live in cities. In addition to the simple increase in population, per person consumption of water has risen. As more people begin utilizing modern luxuries like flush toilets, showers, and washing machines, the demand created by the residential sector will increase dramatically.

Water Pollution Is Getting Worse

The companion of modernization has always been pollution. In developing countries that are just entering the industrial age, water pollution presents a serious problem. According to United Nations Environmental Program (UNEP), "in developing countries, rivers downstream from major cities are little cleaner than open sewers." The UNEP also reports that 1.2 billion people are being affected by polluted water, and that dirty water contributes to 15 million child deaths every year. In recent years, scientists have become aware of the problems involved with the contamination of groundwater. Aquifers move very slowly, so once they are polluted it takes decades or centuries for them to cleanse themselves.

Food production contributes significantly to water contamination. When nitrogen fertilizer is applied to a field, the water runoff will contain excess amounts of nitrates. Nitrates have been shown to have a very harmful effect on plant and animal life, can cause miscarriages, and can harm infant development. The industrial livestock business also presents a serious danger to water systems. The disposal of vast amounts of animal feces destroys nearby ecosystems and is very hazardous to humans.

Water pollution is reaching epic proportions. In the U.S. 40% of rivers and lakes are considered too polluted to support normal activities. In China 80% of the rivers are so polluted that fish cannot survive in them. In Japan 30% of groundwater has been contaminated by industrial pollution. The Ganges River, which supports around 500 million people, is considered one of the most polluted rivers in the world.

Water Shortages Can Lead to Food Scarcity

According to the International Food Policy Research Institute (IFPRI), if current water consumption trends continue, by 2025 the agricultural sector will experience serious water shortages. The IFPRI estimates that crop losses due to water scarcity could be as high as 350 million metric tons per year, slightly more than the entire crop yield of the U.S. This massive water crisis will be caused by water contamination, [the diversion of] water for industrial purposes, and the depletion of aquifers. Climate change may also play a part. The Himalayan glaciers, which feed the rivers that support billions of people, are shrinking in size every year. Their disappearance would cause a major humanitarian disaster.

The greatest danger to global food security comes from aquifer depletion. Aquifers are an essential source of water for food production, and they are being overdrawn in the western U.S., northern Iran, north-central China, India, Mexico, Australia, and numerous other locations: Additionally, many aquifers are contaminated each year by pollution and seawater intrusion.

Despite their importance, data on underground water reservoirs remains imprecise. There is little evidence regarding how many aquifers actually exist, and the depth of known aquifers is often a mystery. However, it is clear that water from these sources takes centuries to replenish, and that they are being consumed at a highly unsustainable rate.

Food production contributes significantly to water contamination because nitrates from fertilizers run off fields and pollute water supplies.

International Conflicts over Water

According to the UNEP, there are 263 rivers in the world that either cross or mark international boundaries. The basins fed by these rivers account for 60% of the world's above ground freshwater. Of these 263 rivers, 158 have no international legislation, and many are the source of conflict.

Water has always been a central issue in the Arab-Israeli situation. Ariel Sharon [former prime minister of Israel] once said the Six Days War [a war fought between Israel and Egypt, Syria and Jordan from June 5–10, 1967] actually began the day that Israel stopped Syria from diverting the Jordan River in 1964. Decades later, the Egyptian military came close to staging a coup against Egyptian president Anwar Sedat, who had proposed diverting some of the Nile's water to Israel as part of a peace plan.

The Nile River, which runs through Ethiopia, Sudan, and Egypt, exemplifies the potential for future water conflicts. The banks of the Nile River support one of the most densely populated areas on the planet. In the next fifty years the number of people dependant on the Nile could double, creating a serious water crisis in the region. The Nile is not governed by any multilateral treaties, and Egypt would not shrink from using military strength to guarantee its future access to water.

The potential for water conflicts [is] less likely outside the Middle East, but nevertheless there are many problematic areas. The Mekong River is the lifeblood of South East Asia, but it begins in one of the most water-poor countries on Earth: China. The Indus River separates Pakistan and India, and . . . [Indian farmers deplete aquifers at] one of the highest rates in the world. U.S.-Mexican relations are already strained over water use on their mutual border. The Niger River basin in West-Central Africa runs through five countries. Surging populations coupled with decreasing rainfall in the region seriously threaten water security for millions of people.

Although the specter of international water wars can seem very real, in the last 50 years there have only been 7 conflicts over water outside the Middle East. While a global water crisis has the potential to tear international relations at the seams, it also has the potential to force the global community into a new spirit of cooperation.

Potential Solutions

The oceans contain 97% of the world's water. Desalination technology transforms the vast amount of salt water in the Earth's oceans into freshwater fit for human consumption. There are approximately 7,500 desalination plants in the world, 60% of which are in the Middle East. The global desalination industry has a capacity of approximately 28 million m^3, less than 1% of global demand. Desalination is an expensive and energy-intensive technology, and currently only wealthy countries with serious water shortages consider it a viable option. However, a recent innovation using nanotechnology has the potential to decrease the cost of desalination by 75%, making it a more viable option.

While irrigation accounts for approximately one-third of all global water consumption, numerous studies have shown that approximately half of the water used in irrigation is lost through evaporation or seepage. Drip irrigation technology offers a far more water-efficient way of farming. Drip irrigation techniques involve using a series of pipes to distribute water in a very controlled manner. By using this method farmers have the ability to give their crops the exact amount of water needed. Despite its many benefits, drip irrigation is not being widely implemented. While the technology is not sophisticated or expensive, it is beyond the means of the poorest farmers who need it most. It is also not being used by many farmers in water-rich countries because the potential savings are less than the cost of implementing the technology.

In many countries water shortages are exacerbated or even caused by governmental mismanagement, political infighting, and outright corruption. International organizations like the World Trade Organization (WTO) often suggest that privatization of water management services would alleviate many of these problems. It has been shown that privatizing utilities frequently increases efficiency, innovation, and maintenance. However, privatization rarely has an effect on corruption, and often disadvantages the poor.

Other technical solutions like rainwater capture, water-free toilets, and water reclamation offer people the possibility of effective conservation. Market-oriented solutions such as water tariffs, pricing groundwater, and increasing fines against industries that pollute could be adopted. There are also a number of viable trade solutions. Freshwater could be traded internationally by using pipelines and enormous plastic bags. Despite this plethora of potential solutions, there is no substitute for simply consuming less.

In the coming decades, water crises will likely become increasingly common. If the population continues to grow at a rate of 1 billion people every 15 years, the Earth's capacity to support human life will be severely strained. Population growth notwithstanding, the current supply of water is being degraded by pollution, overdrawing, and climate change. It is not too late to guarantee a safe supply of water for everyone alive today and for all future generations; although to do so would require an unprecedented level of international cooperation, trust, and compassion.

THREE

Access to Water Is a Human Right

Food and Water Watch

> Food and Water Watch is a nonprofit organization that advocates for commonsense policies that promote healthy, safe food, and access to clean and affordable drinking water. Food and Water Watch advocates for public control of water resources and services, strong conservation measures, and tough regulation of toxic emissions into water. This article argues that the United States should support a United Nations resolution to recognize the human right to water, which is something the United States has failed to do in the past.

Formal recognition of the human right to water by the United Nations is a vital first step to ensure that all people have access to this most basic human need. Yet the United States government has historically opposed this movement. It is time for the administration of Barack Obama to take a stand for human rights and throw its support behind a U.N. resolution that codifies the human right to water.

For more than a decade, water justice groups have been calling for legal recognition of the human right to water at the United Nations (UN)—as well as at national and local levels—in order to ensure access to safe water for billions of people.

Food and Water Watch, "Yes We Can: Why Obama Must Put Human Rights First and Support the Right to Water," July 2010. Copyright © July 2010, Food and Water Watch. Reproduced by permission.

A Filipino buys clean water from a neighbor. In the Philippines 17 million people lack access to clean water. The UN has passed resolutions stating that access to water is a basic human right.

Nearly two billion people live in water-stressed areas and three billion have no running water within a kilometer of their homes. Every eight seconds a child dies of a water-borne disease that would be preventable with access to safe water and adequate sanitation. According to a recent [2009] World Bank report, by 2030, global demand for water will exceed supply by 40 percent.

A UN declaration on the human right to water would give all people equal access to "sufficient, safe, acceptable, physically accessible and affordable water for personal and domestic uses." International norms set by UNICEF [United Nations Children Fund] and the World Health Organization define this as a 20 liter daily minimum (5.2 gallons), increased to 50 liters per day when including bathing and laundry needs.

The Right to Water Is Controversial

In order to protect the quality of life and guide policies of equality and social justice, the UN legally recognizes certain inalienable rights; for example, the rights to food and shelter.

Passing a UN resolution on the right to water would establish the framework for valuing water in the context of human, social and cultural rights, rather than as a commodity. It would set an example for state governments to replicate in national laws—a critical precedent for a world facing population growth, climate change and a growing middle class, all factors that strongly impact water usage.

No one should ever be denied water for basic living needs because of an inability to pay. Without a formal recognition of the right to water, it will continue to be treated as a commodity to be bought, sold, and managed for private gain instead of public good.

It is often incorrectly assumed that recognizing water as a human right forces governments to provide free water to [their] citizens or to the citizens of other countries.

The US Government Has Not Supported the Right to Water

The U.S. has not supported the human right to water, and has a history of voting against social and cultural rights when they conflict with economic interests. . . .

At every World Water Forum—the largest global water event held every three years—a ministerial resolution is produced by consensus. The 2009 forum concluded without consensus because of a language debate over inclusion of the human right to water. Bolivia, Uruguay, Spain and several other countries lobbied strongly for inclusion, only to be blocked by the U.S., Canada, Egypt and Brazil.

US Citizens Support the Right to Water

In contrast to the anti-rights position of the U.S. government in international forums, U.S. municipalities have begun enacting right to water resolutions at the local and state level. In 2006, the Detroit City Council passed a resolution declaring the right to water and preventing water shut-offs for low-income people. Numerous cities and states have passed laws preventing water shut-offs for low-income or elderly residents, particularly during winter months.

Both Massachusetts and Pennsylvania have explicitly recognized the right to water in their state constitutions. In 2009,

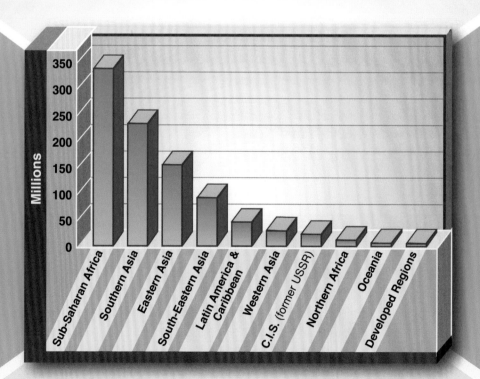

Taken from: World Health Organization, Progress on Sanitation and Drinking Water. Data for 2008.

California passed a right to water bill that Governor Arnold Schwarzenegger vetoed.

The Obama Administration Should Act

While municipalities and states in the U.S. are increasingly moving to pass resolutions recognizing the human right to water, the federal government continues to block them in global forums. It is time for the Obama Administration to break from the unjust and narrow policies of previous administrations and bring its policies into alignment with the American people.

Without support from the U.S. government, the international community is unlikely to reach a consensus on the human right to water. President Obama, who was elected into office on the motto, "Change We Can Believe In," has the opportunity to be a human rights leader by swinging his administration's support behind a resolution. More and more people are recognizing that water is not a commodity, but rather a valuable engine for human development in an increasingly water-stressed world.

While the establishment of water as a human right is necessary and critical, it is also insufficient in providing water for all. Beyond formal recognition, countries must work towards public water systems that are transparently and democratically managed and include citizen participation at all levels of service.

There is absolutely enough water in the world to meet human requirements for health and safety. The only thing missing is the collective political will.

FOUR

Making Water a Human Right Will Not Improve Access

Elizabeth Dickinson

> *Foreign Policy* is a magazine of global politics, economics, and ideas. Elizabeth Dickinson is the assistant managing editor. Prior to joining *Foreign Policy*, Dickinson worked in West Africa as the Nigerian correspondent for the *Economist*. In the following viewpoint she argues that making water a human right, which the United Nations recently did, may hurt poor countries that lack access to clean water. Rather, Dickinson says, putting a price on water would encourage conservation and incentivize private companies to provide water to more people.

The United Nations General Assembly voted yesterday [July 28, 2010] in favor of an international human right to water—something that activists have long been calling for and a few countries (most notably, South Africa) have already instituted. Asserting this unalienable human right, some argue, is the best way to ensure that all citizens, regardless of their location, economic situation, or anything else, have access to what is literally the world's most indispensible commodity.

Elizabeth Dickinson, "Making Water a Human Right May Not Be Such a Good Idea," *Foreign Policy*, July 29, 2010. Copyright ©2010, *Foreign Policy*. Reproduced by permission.

A "Human Right" to Water May Result in Less Access to Water

I'm all for that. But a right to water is not something that can simply be drafted into law, poof! Unfortuantely, getting clean water to every human on the planet is a question of getting water markets and delivery systems up to snuff—and if an universal human right to water was ever adopted, it could actually be a deterrant to that ever happening.

Because nearly everything that creates economic wealth uses water—including individuals, companies, industries, governments, and power plants—it has become clear there must be limits to the amount of water one person or enterprise can use.

Bear with me, I am not a heartless westerner, I promise. Here's the conundrum: All humans need water. But companies, industries, governments, power plants, and nearly everything that creates economic wealth also requires water. And for those entities, water has an economic value. (Water also has an economic value to humans, the way food does.) And like all things that create economic value, we don't have an unlimited supply of water. That means that no matter what, there will limits about how much any one person or business can use.

Putting a Pricetag on Water Will Improve Conservation

Here's where the right complicates things. Rights to something like water usually imply that there is also no price—and often no limits. This is the situation in much of the world today, actually—and in much of the world, there isn't clean water to go around. Elsewhere where the price is too low, for example in the United States, commercial over-use is a real problem. In the U.S. case, underpriced water in the Western states, particularly California, has boosted agricultural use of water and discouraged conservation—to the extent that the state faces an impending water crisis.

Countries that have dramatically improved access to clean water in recent years, on the other hand, include those who have put a pricetag on it—Chile, for example, as well as Britain and Australia. Giving water a price (and certainly one that is bracketed by economic group, with different rates for private and commercial use) encourages conservation and smart use. Moreover, it gives private companies a real incentive to provide the stuff to more consumers. (In some communities, a push—like a subsidy or a tax incentive—would of course be needed.) In the end, all this will help bring more and more people onto the water grid. And that's what it's all about, right?

There Is a Water Crisis in the Middle East

John Bohannon

> John Bohannon is a contributing writer for *Science* magazine and investigates the ethical aspects of science and health policy. He also runs GonzoLabs, a research institution exploring interaction among art, culture, and science. In this article Bohannon discusses the causes and effects of the severe water shortage in Gaza Strip, a territory in Palestine. He argues that the issue is reaching a critical stage and requires immediate attention. Possible solutions include improved wastewater treatment and desalination, as well as a reduction in the population of the area.

You can almost hear the collective sigh of relief as the angry sun sets over this dusty city [Rafah] on Gaza's Egyptian border. This is when five of Ali Abu Taha's sons arrive, unwinding their kaffiyehs and gathering around the charcoal fire where a pot of tea is already boiling. The unprecedented visit of a foreign guest calls for a demonstration of the hospitality for which the Bedouins are famous. The seat of honor is offered, and some of the family's most valuable possessions are laid out on the carpet for display: a battered old AK-47 rifle and several bottles of home-filtered water. The gun is a family heirloom that rarely sees light, but the water is indispensable. "The filter cartridges are very expensive

John Bohannon, "Running Out of Water—and Time," *Science,* August 25, 2006. Copyright © 2006, *Science*. Reproduced by permission.

Palestinian civilians and police officers line up to fill cans with water at a distribution point in Gaza. The water shortage in Gaza is indicative of water problems throughout the Middle East.

and hard to get into Gaza," says one of the sons, Mohammed, "but this one should hold up for another month, *insha' Allah*." Not only does it provide the drinking water for Abu Taha's clan—about 100 people, a third of them his grandchildren—but by enabling them to bottle and sell water to neighbors, it provides one of their few sources of income.

"I don't recommend drinking too much of this," says Mohammed as he fills a glass with unfiltered water from the tap. One sip of the pongy brine is enough to understand why. As a general rule, the farther south one goes in Gaza, the worse the water becomes, and Rafah is the end of the line. The Palestinian

Authority issues warnings from time to time urging the public to buy bottled water, especially for the very young or elderly. But for the average Gazan—with an annual income of $600—a $1 gallon of water is a luxury.

For Abu Taha, who grew up as a Bedouin in the nearby Negev desert, making efficient use of scarce water resources is nothing new. The problem is that the 1.4 million people crammed into the Gaza Strip—most of them the children of refugees who fled their homes in the 1948 and 1967 Arab-Israeli wars—depend on a shallow aquifer for water. But year by year, that source is becoming more contaminated by salt and pollution. Most wells already produce water that is nonpotable [not suitable for drinking] according to standards set by the World Health Organization.

Water scarcity is a perennial problem in the region, but nowhere is it worse than in Gaza. "It is a microcosm of the entire Middle East," says Eric Pallant, an environmental scientist at Allegheny College in Meadville, Pennsylvania, who has collaborated with both Israelis and Palestinians on water problems. "If you can figure out how to make water sustainable there, then you can do it anywhere." Several Gaza water projects have been planned by donor countries in recent years, including state-of-the-art wastewater treatment and desalination plants, but all have fizzled due to security concerns and sanctions slapped onto the new Hamas-led Palestinian government [classified by many governments as a terrorist group]. Israel's withdrawal of settlers and troops from Gaza last year [2005] was a bittersweet victory for the Palestinians. Although they are fully in control of Gaza's water for the first time, they must now scramble to save it before it becomes irreversibly contaminated. . . .

A Severe Water Shortage

At a glance, the Gazans' water woes seem insurmountable. The only natural fresh source available is the coastal aquifer, a soggy sponge of sediment layers that slopes down to the sea a few dozen meters beneath their feet. Its most important input is the meager 20 to 40 centimeters of annual rainfall that sprinkles over

Gaza's 360-square-kilometer surface—about twice the area of Washington, D.C.—giving between 70 and 140 million cubic meters (MCM) of water per year. Most of that water evaporates, but between 20 and 40 MCM penetrates the sandy sediment to feed the aquifer. Another 15 to 35 MCM, depending on whom you ask, flows in under the border from Israel, while irrigation and leaky pipes are estimated to return 40 to 50 MCM, for a total annual recharge of 75 to 125 MCM.

The aquifer's only natural output is the 8 MCM per year that should exit into the Mediterranean, providing a crucial barrier against the intrusion of seawater. So if no more than about 100 MCM were tapped from the aquifer per year, it could last forever. But Gaza's 4000 wells suck out as much as 160 MCM yearly, says Ahmad Al-Yaqoubi, a hydrologist who directs the Palestinian Water Authority. This estimated 60-MCM annual water deficit is why the water table is dropping rapidly and already reaches 13 meters below sea level in some places. Saltwater from the Mediterranean as well as deeper pockets of brine get sucked in to fill the gap. "The saltwater intrusion is well under way," says Al-Yaqoubi, "especially in the coastal areas and to the south." About 90% of wells already have salinity exceeding the WHO-recommended maximum of 250 parts per million (ppm). The accelerating rate of saltwater intrusion alone could make the Gaza aquifer unusable within 2 or 3 decades, according to a 2003 report by the United Nations Environment Programme.

Gaza Faces Sewage Problems as Well

But there may be far less time on the clock. The aquifer is also mixing with a cocktail of pollutants from Gaza's sewage and agriculture. "Besides salt, our number-one contaminant is nitrate from solid waste and fertilizers," says Yousef Abu Safieh, an environmental scientist based in Gaza City who heads the Palestinian Environmental Quality Authority. The maximum safe concentration of nitrate according to WHO is 45 ppm. "In our sampling, we find that most wells have about 200 ppm, and wells close to agricultural runoff can even hit 400," says Abu Safieh. Two Palestinian governmental studies led by Abu Safieh point to patterns of disease matching the distribution of water contamination. The higher the salinity of local water, the higher the incidence of kidney disease, he says, and nitrate concentration correlates with Gaza's high incidence of blue baby syndrome: a loss of available oxygen in the blood that can cause mental retardation or be fatal.

It is the job of a water utility to clean up such contamination and make sure that safe water comes out of the tap, but there is no such unified utility in Gaza. Instead, the strip is covered by a patchwork of fragmented water infrastructure. Gaza's three wastewater treatment plants are far from adequate. The largest, south of Gaza City, was designed to treat 42,000 cubic meters per day— the amount produced by 300,000 people—but now faces a daily inflow of more than 60,000 cubic meters, says Al-Yaqoubi: "This has overwhelmed the biological step of the treatment process." As an emergency measure to prevent sewage from overflowing, barely treated wastewater is now piped to the coast, where the dark gray liquid can be seen, and smelled, flowing along the beach. Meanwhile, the 40% of Gazans without access to a centralized sewage-disposal system contribute to the burgeoning cesspits. A 40-hectare lake of sewage that has formed in northern Gaza is a menace to people at the surface and the aquifer beneath.

These threats to the water supply are serious, says Al-Yaqoubi, but "water scarcity is of course the problem that will never go away." Considering that crop irrigation gobbles up 70% of Gaza's water and fertilizers contribute most of the nitrate contamination, firmer control of agriculture by [the ministry of agriculture] seems like a necessary first step in saving the aquifer from ruin. "The problems continue to spiral," says Mac McKee, a hydrologist at Utah State University in Logan, who has collaborated with Gazans for the past 10 years, because "the Palestinian Authority has not succeeded in applying effective controls on well-drilling and pumping." About half of Gaza's wells have been dug illegally, mostly by farmers to irrigate small plots of cropland.

People Will Not Use Less

"If you try to tell farmers to stop using their wells, they come out with guns," says Ehab Ashour, a water engineer who works for international development agencies in Gaza. And with the struggle for power intensifying between the Hamas and Fatah [a Palestinian political organization] leaderships, the prospect of better enforcement seems dimmer than ever. For his part, [Mohammed Al-Agha, the Hamas minister of agriculture] says a crackdown on well-digging isn't even on the table. "We can't do this from an economic standpoint," he says. "Over 60% of people here are farming. We are all locked into this jail, so we have to grow our own food and at the same time try to produce something we can export."

Asking families in Gaza to use less water is also "out of the question," says David Brooks, an environmental scientist.... Average daily domestic water consumption in Gaza is about 70 liters per person—used not only in homes but also hospitals, schools, businesses, and public institutions—whereas 100 liters per capita per day is the generally agreed minimum for public health and hygiene. (By comparison, average consumption in Israel is 280 liters per day.) So the only way forward is to secure new sources of fresh water and make existing sources stretch farther, says Abu Safieh. There are several strategies for doing this, he says, "and we are pursuing all of them."

Desalination Is a Potential Solution

If you stand on any hill in Gaza and look west, a tantalizing source of water shimmers into view. If only the salts could be efficiently removed, the Mediterranean is a virtually limitless supply for desalination plants. Indeed, this very water is feeding some of the world's most advanced facilities less than an hour's drive up the coast in Israel.

For any long-term solution in Gaza, "desalination will be absolutely necessary," says McKee. A desalination plant capable of providing Gaza with 60 MCM of drinking water per year was part of a plan drawn up by the United States Agency for International Development (USAID) in 2000. Money to build the $70 million plant, along with $60 million to lay down a carrier system to pipe the water across Gaza, was ready to go from USAID when the second intifada [the second Palestinian uprising] broke out just months

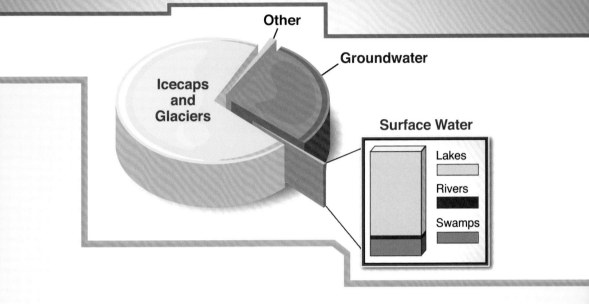

Taken from: P.H. Gleick, "1996: Water Resources." In *Encyclopedia of Climate and Weather*, ed. by S.H. Schneider, Oxford University Press, New York, vol. 2, pp. 817–823.

later, stalling the project. It was officially frozen in 2003 after a bombing killed three members of a U.S. diplomatic convoy in Gaza.

Besides producing more drinking water, the priority is to deal with Gaza's sewage, says Al-Yaqoubi, not only to prevent a public health disaster but also to recycle some of the precious water back into the system. A trio of wastewater treatment plants that could handle Gaza's entire load has been promised by USAID, the World Bank, Germany, Finland, and Japan, but "nothing has happened," he says, because of the Hamas election victory.

In relation to stretching the current water supply farther, there is one positive legacy of Israeli occupation in Gaza. By working on Israeli farms, "we have become very comfortable with new technologies," says Al-Yaqoubi. In spite of the official freeze on international aid to the Palestinian government, projects aiming to improve farming in Gaza "are ongoing by many donors," he says. The most important is drip irrigation, delivering water directly to roots through a network of tubes. Coupling this with a computerized system that automatically pumps just enough water from a well to meet the plants' daily needs can make irrigation up to 70% more efficient over the long run.

Help from Other Countries

But for the immediate crisis, the country best placed to help Gaza may be Israel. Before the taps were shut this year [2006] after Hamas was elected, 5 MCM per year of drinking water was being piped into Gaza by Mekorot, the Israeli national water company, and an additional 5 MCM had been agreed. That water does not come free, but it is nevertheless a freshwater source separate from the ailing aquifer.

"We know how serious the situation is in Gaza," says Saul Arlosoroff, a member of Mekorot's board of directors and a former Israeli deputy water commissioner. "The first priority is to get these people enough clean drinking water, and the second is to prevent salinity from irreversibly destroying their soil." Arlosoroff says Israelis and Palestinians working in the water sector have a special relationship. "We understand each other, and we know that these problems require cooperation," he says, "but the atmosphere between Gaza and Israel is worse now than at any time in our history."

Across the border, Abu Safieh is similarly disappointed. "There was a time when I could talk with my Israeli counterpart constructively about our environmental problems," he says, but he has not had any contact in years. Al-Agha says he plans to turn to Egypt for help. For importing and exporting, as well as perhaps for obtaining the abundant electricity needed to desalinate water, he says, "our hope is to the south."

The present turmoil also prevents what Brooks calls "the easiest and best solution" to Gaza's environmental problems: reducing the number of people living there. "Gaza can't sustain that population, and any real solution will require people to leave," he says. Most Gazans "will never give up hope of returning to their homes," says Abu Safieh, but for now, "we will work to make the best of the bad situation."

SIX

Africa Suffers from a Water Crisis

Tina Rosenberg

> Tina Rosenberg writes for *National Geographic* magazine. In this story she details the lives of women in the Ethiopian village of Foro. These women spend enormous amounts of time and energy hauling unsanitary water to their village. The water is unsanitary, but it is the only water available. Rosenberg explains that through education and technology, groups such as WaterAid are improving the lives of the people in Foro and similar villages.

Aylito Binayo's feet know the mountain. Even at four in the morning she can run down the rocks to the river by starlight alone and climb the steep mountain back up to her village with 50 pounds of water on her back. She has made this journey three times a day for nearly all her 25 years. So has every other woman in her village of Foro, in the Konso district of southwestern Ethiopia. Binayo dropped out of school when she was eight years old, in part because she had to help her mother fetch water from the Toiro River. The water is dirty and unsafe to drink; every year that the ongoing drought continues, the once mighty river grows more exhausted. But it is the only water Foro has ever had.

The task of fetching water defines life for Binayo. She must also help her husband grow cassava and beans in their fields, gather

---

Tina Rosenberg, "The Burden of Thirst," *National Geographic*, April 2010. Copyright © 2010, National Geographic Society. Reproduced by permission.

grass for their goats, dry grain and take it to the mill for grinding into flour, cook meals, keep the family compound clean, and take care of her three small sons. None of these jobs is as important or as consuming as the eight hours or so she spends each day fetching water.

In wealthy parts of the world, people turn on a faucet and out pours abundant, clean water. Yet nearly 900 million people in the world have no access to clean water, and 2.5 billion people have no safe way to dispose of human waste—many defecate in open fields or near the same rivers they drink from. Dirty water and lack of a toilet and proper hygiene kill 3.3 million people around the world annually, most of them children under age five. Here in southern Ethiopia, and in northern Kenya, a lack of rain over the past few years has made even dirty water elusive. . . .

Lack of Water Leads to Poor Hygiene

When you spend hours hauling water long distances, you measure every drop. The average American uses a hundred gallons of water just at home every day; Aylito Binayo makes do with two and a half gallons. Persuading people to use their water for washing is far more difficult when that water is carried up a mountain. And yet sanitation and hygiene matter—proper hand washing alone can cut diarrheal diseases by some 45 percent. Binayo washes her hands with water "maybe once a day," she says. She washes clothes once a year. "We don't even have enough water for drinking—how can we wash our clothes?" she says. She washes her own body only occasionally. A 2007 survey found that not a single Konso household had water with soap or ash (a decent cleanser) near their latrines to wash their hands. Binayo's family recently dug a latrine but cannot afford to buy soap.

Much of the cash they do have goes for four- to eight-dollar visits to the village health clinic to cure the boys of diarrhea caused by bacteria and parasites they regularly get from the lack of proper hygiene and sanitation and from drinking untreated river water. At the clinic, nurse Israel Estiphanos said that in normal times 70 percent of his patients suffer waterborne diseases—and

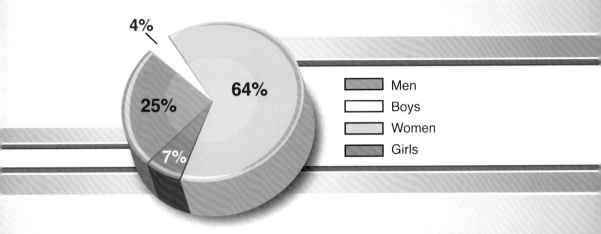

Taken from: United Nations, *The Millennium Development Goals Report, 2008*. Data from 2005/2006.

now the area was in the midst of a particularly severe outbreak. Next to the clinic a white tent had arisen for these patients. By my next visit, Estiphanos was attending to his patients wearing high rubber boots.

Sixteen miles away at the district health center in Konso's capital, almost half the 500 patients treated daily were sick with waterborne diseases. Yet the health center itself lacked clean water. On the walls of the staff rooms were posters listing the principles of infection control. But for four months a year, the water feeding their taps would run out, said Birhane Borale, the head nurse, so the government would truck in river water. "We use water then only to give to patients to drink or swallow medicine," he said. "We have HIV patients and hepatitis B patients. They are bleeding, and these diseases are easily transmittable—we need water to disinfect. But we can clean rooms only once a month."

Even medical personnel weren't in the habit of washing hands between patients, as working taps existed at only a few points in the building—most of the examining rooms had taps, but they were not connected. Tsega Hagos, a nurse, said she had gotten spattered with blood taking out a patient's IV. But even though there was water that day, she had not washed her hands afterward. "I just put on a different glove," she said. "I wash my hands when I get home after work."

The Challenges of Accessing Clean Water

Bringing clean water close to people's homes is key to reversing the cycle of misery. Communities where clean water becomes accessible and plentiful are transformed. All the hours previously spent hauling water can be used to grow more food, raise more animals, or even start income-producing businesses. Families no longer drink microbe soup, so they spend less time sick or caring for loved ones stricken with waterborne diseases. Most important, freedom from water slavery means girls can go to school and choose a better life. The need to fetch water for the family, or to take care of younger siblings while their mother goes, is the main reason very few women in Konso have attended school. Binayo is one of only a handful of women I met who even know how old they are.

Access to water is not solely a rural problem. All over the developing world, many urban slum dwellers spend much of the day waiting in line at a pump. But the challenges of bringing water to remote villages like those in Konso are overwhelming. Binayo's village of Foro sits atop a mountain. Many villages in the tropics were built high in the hills, where it is cooler and less malarious and easier to see when the enemy is coming. But Konso's mountaintop villages do not have easy access to water. Drought and deforestation keep pushing the water table lower—in some parts of Konso it is more than 400 feet belowground. The best that can be done in some villages is to put in a well near the river. The water is no closer, but at least it is reliable, easier to extract, and more likely to be clean.

Yet in many poor nations, vast numbers of villages where wells are feasible do not have them. Boring deep holes requires geological know-how and expensive heavy machinery. Water in many countries, as in Ethiopia, is the responsibility of each district, and these local governments have little expertise or money. "People who live in slums and rural areas with no access to drinking water are the same people who don't have access to politicians," says Paul Faeth, president of Global Water Challenge, a consortium of 24 nongovernmental groups that's based in Washington, D.C. So the effort to make clean water accessible falls largely to charity groups, with mixed success.

The villages of Konso are littered with the ghosts of water projects past. In Konsos around the developing world, the biggest problem with water schemes is that about half of them fall into disrepair soon after the groups that built them move on. Sometimes technology is used that can't be repaired locally, or spare parts are available only in the capital. But other reasons are achingly trivial: The villagers can't raise money for a three-dollar part or don't trust anyone to make the purchase with their pooled funds. The 2007 survey of Konso found that only nine projects out of 35 built were functioning.

Charities Are Helping

Today [2010] a U.K.-based international nonprofit organization called WaterAid, one of the world's largest water-and-sanitation charities, is tackling the job of bringing water to the most forgotten villages of Konso. At the time of my visit, WaterAid had repaired five projects and set up committees in those villages to manage them, and it was working to revive three others. At the health center in Konso's capital, it was installing gutters on the sloped roofs of the buildings to conduct rainwater to a covered tank. The water is now being treated and used in the health center.

WaterAid is also working in villages like Foro, where no one has successfully brought water before. Their approach combines technologies proven to last—such as building a sand dam to capture and filter rainwater that would otherwise drain away—with new ideas like installing toilets that also generate methane gas for a new communal kitchen. But the real innovation is that WaterAid treats technology as only part of the solution. Just as important is involving the local community in designing, building, and maintaining new water projects. Before beginning any project, WaterAid asks the community to form a WASH (water, sanitation, hygiene) committee of seven people—four of whom must be women. The committee works with WaterAid to plan projects and involve the village in construction. Then it maintains and runs the project.

The people of Konso, who grow their crops on terraces they have painstakingly dug into the sides of mountains, are famous for hard work, and they are an asset—one of Konso's few—in the quest for water. In the village of Orbesho, residents even built a road themselves so that drilling machinery could come in. Last summer their pump, installed by the river, was being motorized to push its water to a newly built reservoir on top of a nearby

Ethiopian women carry water from a lake. Because of water shortages caused by a severe drought, Ethiopians have been forced to draw water from unsanitary sources.

mountain. From there, gravity would pipe it down to villages on the other side of the mountain. Residents of those villages had contributed a few cents apiece to help fund the project, made concrete, and collected stones for the structures, and now they were digging trenches to lay pipes. . . .

Changing Habits to Improve Hygiene

If installing a water pump is technically challenging, encouraging hygiene is a challenge of a different kind. Wako Lemeta is one of two hygiene promoters whom WaterAid has trained in Foro. Lemeta, rather shy and poker-faced, stops by Binayo's house and asks her husband, Guyo Jalto, if he can check their jerry cans. Jalto leads him to the hut where they are stored, and Lemeta uncovers one and sniffs. He nods approvingly; the family is using WaterGuard, a capful of which purifies a jerry can of drinking water. The government began to hand out WaterGuard at the beginning of the recent outbreak of disease. Lemeta also checks if the family has a latrine and talks to villagers about the advantages of boiling drinking water, hand washing, and bathing twice a week.

Many people have embraced the new practices. Surveys say latrine use has risen from 6 to 25 percent in the area since WaterAid began work in December 2007. But it is a struggle. "When I tell them to use soap," Lemeta explains, "they usually tell me, 'Give me the money to buy it.'"

Similar barriers must be overcome to keep a program going after the aid group leaves. WaterAid and other successful groups, such as Water.org, CARE, and A Glimmer of Hope, believe that charging user fees—usually a penny per jerry can or less—is key to sustaining a project. The village WASH committee holds the proceeds to pay for spare parts and repairs. But villagers think of water as a gift from God. Should we next pay to breathe air?

Cost Is an Issue

Water and money have long been an uneasy mixture. Notoriously, in 1999 Bolivia granted a multinational consortium 40-year rights to provide water and sanitation services to the city of

Cochabamba. The ensuing protests over high prices eventually drove out the company and brought global attention to the problems of water privatization. Multinational companies brought in to run public water systems for profit have little incentive to hook up faraway rural households or price water so it is affordable to the poor.

Yet someone has to pay for water. Although water springs from the earth, pipes and pumps, alas, do not. This is why even public utilities charge users for water. And water is often most expensive to provide for those who can least afford it—people in the remote, sparsely populated, drought-stricken villages of the world.

"The key question is, Who decides?" says Global Water Challenge's Faeth. "In Cochabamba nobody was talking to the very poorest. The process was not open to the public." A pump in a rural village, he says, is a different story. "At the local level there is a more direct connection between the people implementing the program and the people getting access to water."

The Konso villagers, for instance, own and control their pumps. Elected committees set fees, which cover maintenance. No one seeks to recoup the installation costs or to make a profit. Villagers told me that, after a few weeks, they realized paying a penny per jerry can is actually cheap, far less than what they were paying through the hours spent hauling water—and the time, money, and lives lost to disease.

How would Aylito Binayo's life be different if she never had to go to the river for water again? Deep in a gorge far from Foro, there is a well. It is 400 feet deep. During my visit it was nothing much to look at—aboveground it was only a concrete box with a jerry can inverted over it for protection, surrounded by a pyramid of bramble bushes. But here's what was to happen by March: A motorized pump would push the water up the mountain to a reservoir. Then gravity would carry it back down to taps in local villages—including Foro. The village would have two community taps and a shower house for bathing. If all went well, Aylito Binayo would have a faucet with safe water just a three-minute stroll from her front door.

Scientific Advances Take the Salt out of Salt Water

Robert F. Service

> Robert F. Service is a staff writer for *Science* magazine. He writes about materials science, chemistry, and scientific institutions in his region of the country, the Pacific Northwest. In the following viewpoint Service discusses scientific breakthroughs in desalination technologies, the process in which salt water is altered to become drinkable freshwater. Because the cost, energy consumption, and environmental impact of desalination are all decreasing, Service argues that desalination should be considered as a viable solution to water shortages around the world.

Efforts to provide clean, fresh water for the world's inhabitants seem to be moving in the wrong direction. According to the World Health Organization, 1 billion people do not have access to clean, piped water. A World Resources Institute analysis adds that 2.3 billion people—41% of Earth's populations—live in water-stressed areas, a number expected to climb to 3.5 billion by 2025. To make matters worse, global population is rising by 80 million a year, and with it the demand for new sources of fresh water.

Robert F. Service, "Desalination Freshens Up," *Science*, vol. 313, August 25, 2006. Copyright © 2006, Science Magazine. Reproduced by permission.

Wealthy countries are by no means immune. In arid parts of the United States and many other countries, groundwater resources are already dwindling, and supplies that remain are becoming increasingly brackish [somewhat salty]. Environmental concerns have drastically limited the building of new dams in recent decades. In many areas, "we are already wringing all the water out of the systems that they have," says Thomas Hinkebein, a geochemist at Sandia National Laboratories in Albuquerque, New Mexico. "[We] have to start developing new sources of water."

Such concerns have made desalination—the process of removing salts and suspended solids from brackish water and seawater—a fast-growing alternative. According to a 2004 report by the U.S. National Research Council, more than 15,000 desalination plants now operate in more than 125 countries, with a total capacity of turning out 32.4 million cubic meters (m^3) of water a day, about one-quarter of the amount consumed by U.S. communities each year. With numerous areas around the globe facing long-term severe water shortages, "I don't see [the demand for desalination] slowing down any," says Michelle Chapman, a physical scientist at the U.S. Bureau of Reclamation in Denver, Colorado, and co-chair of a desalination research program funded by the U.S. Office of Naval Research.

But desalination faces its own problems. The two technologies at the heart of conventional desalination plants—evaporation and reverse osmosis (RO), which involves pushing water through a semipermeable membrane that blocks dissolved salts—both require huge amounts of energy. A typical seawater RO plant, for example, requires 1.5 to 2.5 kilowatt-hours (kWh) of electricity to produce 1 m^3 of water; a thermal distillation plant sucks up to 10 times that amount. Countries such as Saudi Arabia may be able to afford to run such facilities, but for most other countries, the cost was already too high even before oil prices went through the roof.

Yet despite those worrisome trends, the prospects for desalination have brightened considerably in the past few years. New engineering designs have slashed the cost of desalination plants, particularly membrane-based RO systems, and new technologies as diverse as nanotechnology and novel polymers are expected to

Israel's Hadera desalination plant is the largest in the world. It uses the latest reverse osmosis technology to extract freshwater from seawater.

drive down operating costs in the years ahead. "There is a huge body of research going on," says Hinkebein, who also oversees a broad collaboration on charting a future road map for desalination technology. "Progress has been a bit incremental for a number of years," adds Anne Mayes, a materials scientist and membrane specialist at the Massachusetts Institute of Technology (MIT) in Cambridge. "But now new opportunities are starting to open up. We're going to see some very different technologies being developed in the near future."

Faster, Cheaper, Better

Desalination has ancient roots. Aristotle and Hippocrates described the process of evaporating salt water to make fresh water in the 4th century B.C.E. In modern times, desalination kicked into gear in the early 20th century. By the mid-1950s, hundreds of desalination plants were on line. Most were based on evaporation, a technique that continues to turn out about half of the globe's desalinated water. Although typically more expensive, the technique remains popular in the Middle East, largely because it is well-suited for dealing with the high levels of salts and suspended solids in the water of the Persian Gulf.

Elsewhere, most new plants being built today use RO because the process requires far less energy. As its name implies, the technology reverses the process of osmosis: the natural tendency of water molecules to flow through a semipermeable membrane to dilute a chemical solution on the other side, in this case seawater. To force water molecules to travel the other way requires pressure—at least 3 megapascals (MPa), but more typically 6 MPa—which in turn requires electricity. Historically, an RO plant has used 10 to 15 kWh of electricity to produce 1 m^3 of fresh water.

Between 1980 and 2000, improvements in pumps and other equipment in RO plants dropped the amount of energy needed to produce fresh water by about half, says John MacHarg, CEO of the Affordable Desalination Coalition (ADC), a San Leandro, California–based group of 22 municipalities, state agencies, and desalination companies looking to improve seawater desalination technology. Since 2000, energy requirements have again dropped by about half, thanks to new energy-recovery devices called isobaric chambers that redirect pressure from the waste brine to low-pressure incoming water. These devices recover up to 97% of the energy. Their resounding success has already made them an integral part of the newly designed desalination plants. In one such plant, which started up last fall in Ashkelon, Israel, for example, isobaric chambers have helped lower the cost of desalinated water to $0.527 cents per m^3, among the cheapest ever by a desalination facility.

Such price drops are now widely expected to continue. By combining energy-recovery devices with new low-pressure membranes and other commercially available advances, this spring ADC members set a new world record for low-cost desalination, dropping the energy needed to 1.58 kWh per m³ of water produced. At that rate, a seawater desalination plant could supply a typical U.S. household with fresh water for the amount of power needed to light an 80-watt light bulb, MacHarg says. That figure, he adds, could change the equation of how to supply places such as southern California with water, because it takes the same amount of energy to pump freshwater from northern California to Los Angeles. . . .

New Advances

Other low-energy desalination techniques are also on the horizon. In one, called forward osmosis, researchers try to harness normal osmotic pressure for making freshwater. They start with freshwater and seawater separated by a membrane and spike the freshwater side with a high concentration of sugar. Freshwater flows through the membrane as it works to dilute the high sugar concentration. "The problem is that you end up with sweetened water," says Menachem Elimelech, an environmental engineer at Yale University. In place of sugar, Elimelech and colleagues have been experimenting with dissolved ammonium salts, such as ammonium bicarbonate. The salts draw fresh water through the membrane without the need for added pressure. Then, by heating the solution to 58°C, Elimelech's team causes the dissolved salts to form ammonia and carbon dioxide gases, which are easily separated from the water. "If we can use waste heat, the process can be very economical," Elimelech says.

Two other technologies are also looking to waste heat and very cheap starting materials to make easily affordable desalination systems. One, dubbed "dewvaporation," is the brainchild of James Beckman, a chemical engineer at Arizona State University, Tempe. The other, called membrane distillation, has been pioneered by Kamalesh Sirkar, a chemical engineer at the New Jersey Institute of Technology in Newark. Beckman's dewvaporation apparatus vaporizes water in one compartment, sending it over a barrier to another where it condenses; Sirkar's membrane distillation passes the water vapor through pores in a membrane

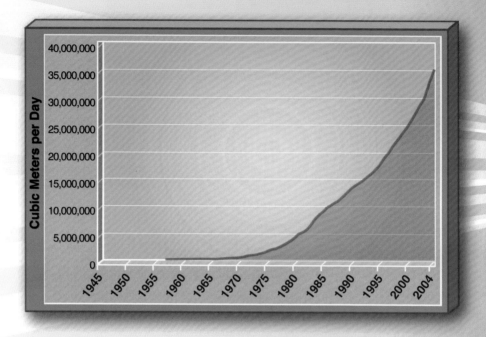

Taken from: Wangnick/GWI, 2005. *2004 Worldwide Desalting Plants Inventory.* Global Water Intelligence. Oxford, England.

that liquid water or larger ions cannot traverse. Both processes are well on their way to proving themselves in the real world. Sirkar's membrane-distillation system is now being put through its paces by United Technologies in East Hartford, Connecticut, and dewvaporation is being evaluated as an option to create freshwater by the city of Phoenix, Arizona.

Beckman and Sirkar say the advantage of their systems is that they can work with a variety of waste-heat sources, such as steam from industrial plants or even solar energy. That versatility could make them especially advantageous for developing countries. Chapman notes that such systems can be particularly useful as add-ons to conventional RO systems. RO plants typically convert only about 50% to 70% of salt water to fresh water and must treat and dispose of the waste brine—a costly process. Because these novel systems can potentially evaporate all the water and leave only solid salts behind, they promise to save governments a lot of money, Chapman says.

It's unclear whether such novel systems will be able to compete with industrial-scale RO and thermal desalination plants. But Chapman points out that the needs of different communities vary widely when it comes to water, depending on the quality of the water source among other factors. "All water sources are different," Chapman says. "So there will probably be a place for all of these technologies"—and no doubt plenty of thirsty users as well.

EIGHT

The Lack of Adequate Sewage Systems Causes Major Health Problems

Barbara Frost

> Barbara Frost is the chief executive of WaterAid, a leading independent organization that enables the world's poorest people to gain access to safe water, sanitation, and hygiene education. Here Frost argues that the lack of proper sanitation and toilets is a global crisis. She argues that much more needs to be done to address the problem and suggests low-cost sanitation solutions that could have significant benefits.

While carbon emissions are now rightly lodged at the forefront of most people's minds, I'm constantly dismayed at how few mentions another urgent global environmental crisis is getting. There is a major stink out there.

It's not a glamorous issue, far from it. In fact it's a bit of a taboo subject. But it's an issue we need to start talking about because it's causing extensive environmental damage and leading to the deaths of 4,000 infants every day. It's a problem of disposal of human waste and toilets. Or rather the lack of good sanitation and toilets.

Much of the World Lacks Adequate Sanitation

Roughly 2.6 billion people—40% of the world's population—live without access to adequate sanitation. Their excrement pollutes

Barbara Frost, "Time to Wake Up to the Sewage Crisis," *The Daily Telegraph*, January 21, 2008. Copyright © 2008, *The Daily Telegraph*. Reproduced by permission.

urban and rural environments, contaminates drinking water supplies with deadly diarrhoeal diseases, reduces oxygen levels in rivers so that plant and animal life dies and causes sewage to build up along coastal fringes.

Wherever I travel in the developing world I meet people who suffer as a result. And it goes without saying it's the poorest people who suffer the most.

People like Bundaa Joseph from the Tabora Region of Tanzania. Just ten years old, Joseph visits a muddy pool every day to fetch water for his family. Due to a lack of sanitation this source is contaminated with raw sewage, as Joseph knows: "I'm not happy to use this water. Some people use it like a toilet."

Even those who have a toilet at home can contribute to extensive pollution if their municipal authorities discharge untreated wastewater and sewage into rivers or seas, as all too many do.

As populations grow, so does the amount of sewage. Levels of suspended solids in Asia's rivers have almost quadrupled since the late 1970s, and indicators show sewage levels in Asia's rivers are 50 times higher than the WHO [World Health Organization] guidelines.

The 'sewage crisis' is inextricably intertwined with the world's freshwater crisis. As climate change affects rainfall patterns and depletes water tables, the amount of water available is dwindling in many areas, a problem exacerbated by increased competition for it from growing populations, agriculture and industry.

It's an untenable situation for the international community to continue to stand by and watch as precious freshwater ecosystems and marine environments are polluted in a way that is easily avoided.

More Needs to Be Done to Address the Crisis

So what's happening to rectify the situation? The answer, sadly, is not nearly enough.

At the Johannesburg World Summit on Sustainable Development in 2002, a Millennium Development Goal (MDG) target was agreed to halve by 2015 the proportion of people without sanitation.

The price tag for solving the problem—estimated at an additional [$10 billion] a year to achieve the MDG target—is about a third of the costs it imposes in healthcare and lost productivity alone. And it's an investment estimated to bring returns of about $9 for each $1 invested.

A Tanzanian listens as an exhibitor explains a new composting toilet, which breaks down human waste to produce a rich, environmentally friendly fertilizer that can boost agricultural yields.

Yet despite the MDG framework and the economic argument, when it comes to sanitation, mostly what we hear from the world's decision-makers is deafening silence.

If current trends of investment in this forgotten and marginalised sector persist the MDG target will be missed by half a billion people.

To help wake the international community up to the severity of the situation, the UN [United Nations] has declared 2008 as the International Year of Sanitation with the aim of trying to get the MDG target back on track.

Many Solutions Are Simple and Cheap

At WaterAid we are fully backing the Year. We'll be using every opportunity possible to demonstrate that establishing safe sanitation need not be an expensive business. It doesn't have to involve the construction of miles of sewers and sophisticated sewage treatment works.

As long as care is taken to build them above the water table, simple pit latrines that cost just a few pounds, and sometimes pence, are perfectly adequate for the disposal of human excreta. All that is needed is an adequately sealed space where excreta can safely decompose, coupled with hygiene education to ensure that communities understand the importance of using them safely.

As well as preventing pollution some low-cost sanitation solutions actively benefit the environment, as WaterAid's widespread use of composting latrines demonstrates. These latrines . . . break down human waste to produce a rich, environmentally friendly fertiliser that can naturally boost agricultural yields.

Rideana Juma from the rural village of Kitayita in the Wakiso District of Uganda is very pleased with her new composting [toilet]. It means her family's excrement no longer pollutes the nearby spring when the rains come. And it has other benefits for her too:

> "The soil around the compound is clean. We can use the products of the latrine for composting. I hope to use the products to help grow bananas and coffee which I will be able to sell. Now I have these facilities I feel more privacy as a lady."

The chief barrier to progress in building and using latrines such as Juma's isn't the cost of the toilets themselves: it's lack of knowledge amongst poor communities on the connection between poor sanitation and disease and how to build latrines.

WaterAid's Community Led Total Sanitation initiatives have shown that when communities learn about the link and calculate how much poor sanitation is costing them in medical bills and lost productivity, they are more than happy to make the modest investment in a basic loo [toilet] themselves. They know they'll see the money back soon enough.

In the rural Rajshahi district of Bangladesh even the kids are now loudly singing the praises of improved sanitation. "The envi-

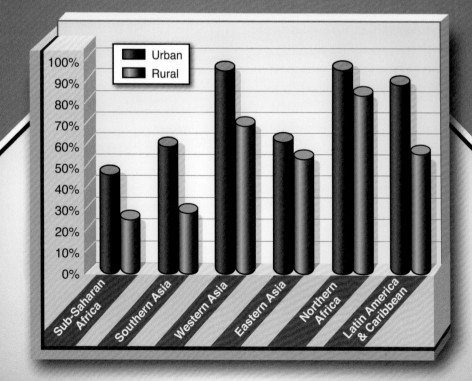

Taken from: United Nations Millenium Development Goals Report, 2010. Data from 2008.

ronment is clean so we can go and play, skipping and running. I feel bad for other villages," said 10-year-old Monira from the village of Laloich.

The main barrier to progress is a lack of political will amongst donors and governments to investing in improving communities' knowledge on sanitation.

The End Water Poverty coalition, of which WaterAid is a founding member, is calling for a global action plan on sanitation, a global taskforce to oversee it and commitments to finance these. The G8 summit in Japan in [2008] is the next big milestone in the international community's decision-making calendar. Let's hope they sit up and listen to these calls.

It is time for action to address the stink and to improve health, dignity and respect for all those without toilets.

NINE

Recycling Sewage Water Is Clean and Effective

Eilene Zimmerman

Eilene Zimmerman is a San Diego–based journalist who writes about business and political and environmental issues. Her work appears in the *New York Times*, the *San Francisco Chronicle*, *Fortune Small Business*, Salon.com, *Wired*, the *Christian Science Monitor*, and other publications. In the following viewpoint Zimmerman argues in favor of new programs to treat and recycle sewage water. Places such as California need new sources of water as old ones are depleted, and such treatment is cheaper than desalination, she says. She explains that the "yuck factor" opposition to sewage recycling is unjustified and that recycled sewage water is at least as pure as water from other sources.

Officials in Orange County, Calif., will attend opening ceremonies [January 2008] for the world's largest water-purification project, among the first "toilet-to-tap" systems in America. The Groundwater Replenishment System is designed to take sewage water straight from bathrooms in places like Costa Mesa, Fullerton, and Newport Beach and—after an initial cleansing treatment—send it through $490 million worth of pipes, filters, and tanks for purification. The water then flows into lakes in nearby Anaheim,

Eilene Zimmerman, "It's Time to Drink Toilet Water," Slate.com, January 25, 2008. Copyright © 2008, Slate.com. Reproduced by permission.

where it seeps through clay, sand, and rock into aquifers in the groundwater basin. Months later, it will travel back into the homes of half a million Orange County residents, through their kitchen taps and showerheads.

It's a smart idea, one of the most reliable and affordable hedges against water shortages, and it's not new. For decades, cities throughout the United States have used recycled wastewater for nonpotable needs, like agriculture and landscaping; because the technology already exists, the move to potable uses seems a no-brainer. But the Orange County project is the exception. Studies show that the public hasn't yet warmed to the notion of indirect potable reuse (IPR)—or "toilet-to-tap," as its opponents would have it. Surveys like one taken last year in San Diego show that a majority of us don't want to drink water that once had poop in it, even if it's been cleaned and purified. A public outcry against toilet-to-tap in 2000 forced the city of Los Angeles to shut down a $55 million project that would have provided enough water for 120,000 homes. Similar reluctance among San Diego residents led Mayor Jerry Sanders to veto the city council's approval in November [2007] of a pilot program to use recycled water to supplement that city's drinking water. (A similar plan failed once before in 1999.)

But San Diego is in the midst of a severe water crisis. The city imports 90 percent of its water, much of that from the Colorado River, which is drying up. The recent legal decision to protect the ecosystem of the San Joaquin Delta in Northern California—San Diego's second-leading water source—will reduce the amount coming from there as well. Add to that rising population and an ongoing drought, and the situation looks pretty bleak: 3 million people in a region that has enough water, right now, for 10 percent of them.

We don't have enough water where we need it; if we don't learn to deal with drinking toilet water, we're going to be mighty thirsty. Only 2.5 percent of the water on Earth is freshwater, and less than 1 percent of that is usable and renewable. The Ogallala Aquifer—North America's largest, stretching from Texas to South Dakota—is steadily being depleted. And Americans are insatiable water consumers—our water footprint has been estimated to be twice the global average.

Desalination Is Not as Efficient

The ocean provides another source of potable water. Large-scale treatment of seawater already occurs in the Middle East, Africa, and in Tampa Bay, Fla. Construction of the largest desalination plant in the western hemisphere is supposed to begin this year in Carlsbad, Calif., which would convert 300 million gallons of seawater into 50 million gallons of drinking water each day. Taking the salt out of ocean water sounds like a good idea, but it's economically and environmentally far more expensive than sewage-water recycling. Orange County water officials estimate desalinated water costs between $800 and $2,000 per acre-foot to produce, while its recycled water runs about $525 per acre-foot. Desalination also uses more energy (and thus produces more greenhouse gas emissions), kills tiny marine organisms that get sucked up into the processing plant, and produces a brine byproduct laced with chemicals that goes back into the ocean.

What desalination doesn't have, though, is the "yuck" factor of recycled sewage water. But seawater, like other sources of non-recycled water, is at least as yucky as whatever comes through a toilet-to-tap program. When you know how dirty all this water is before treatment, recycling raw sewage doesn't seem like a bad option. Hundreds of millions of tons of sewage are dumped into rivers and oceans, and in that waste are bacteria, hormones, and pharmaceuticals. Runoff from rainwater, watering lawns, or emptying pools is the worst, sending metals, pesticides, and pathogens into lakes, rivers, and the ocean. The water you find near the end of a river system like the Colorado or the Mississippi (which feeds big cities like San Diego and New Orleans) has been in and out of municipal sewers several times.

Recycled Water Has Been Shown to Be Safe

Whatever winds up in lakes and rivers used for drinking is cleaned and disinfected along with the rest of our water supply. Still, a

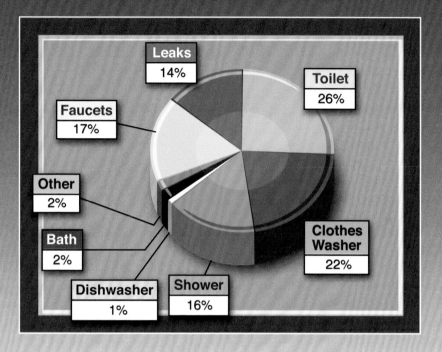

Indoor Home Water Use

- Leaks 14%
- Toilet 26%
- Faucets 17%
- Other 2%
- Bath 2%
- Dishwasher 1%
- Shower 16%
- Clothes Washer 22%

Taken from: National Association of Home Builders.

recent analysis of San Diego's drinking water found several contaminants, including ibuprofen, the bug repellent DEET, and the anti-anxiety drug meprobamate. No treatment system will ever be 100-percent reliable, and skeptics who worry that pathogens in sewage water will make it past treatment and into our drinking water should worry about all drinking water, not just the water in a toilet-to-tap program. The fact is, supertreated wastewater is clean enough to drink right after treatment. It's been used safely this way (in a process known as direct potable reuse) for years in the African nation of Namibia. The EPA [Environmental Protection Agency] has conducted research in Denver and San Diego on the safety of direct potable reuse and found recycled water is often of better quality than existing drinking water. And although putting water into the ground, rivers, or lakes provides

Israel's largest reservoir, the Negev Desert's Besor Reservoir, contains recycled, purified sewage and rainwater used for agricultural purposes.

some additional filtering and more opportunities for monitoring quality, the benefits of doing it that way are largely psychological. In its 2004 report on the topic, the EPA concluded that Americans perceive this water to be "laundered" as it moves through the ground or other bodies of water, even though in some instances, according to the report, "quality may actually be degraded as it passes through the environment."

Despite the public's concerns, a few U.S. cities have already started to use recycled wastewater to augment drinking water. In El Paso, Texas, indirect potable reuse supplies 40 percent of the city's drinking water; in Fairfax, Va., it supplies 5 percent. Unless we discover a new source of clean, potable water, we're going to have to consider projects like these to make wastewater a reusable resource. The upfront costs for getting a system in place and educating the public may be steep, but it would save us the expense—both economic and environmental—of finding another river or lake from which we can divert water.

Bottled Water Is a Huge Industry That Continues to Grow

IBISWorld

> Founded in 1972, IBISWorld provides an extensive online portfolio of business research and analysis products designed to serve a range of business, professional service, and governmental organizations. In addition, the company provides databases of economic analysis, demographic data, and risk-assessment reports. The following report states that the bottled water industry continues to expand at a high rate, particularly the "premium" and "functional" waters. Concerns for the industry include safety monitoring and testing, as well as difficulties faced when expanding into developing countries.

Driven largely by health consciousness consumers with strong disposable income, and warm weather, the bottled water market has been identified by IBISWorld as being the fastest growing beverage segment in the U.S., with the market share for bottled water increasing from 11.7 percent in 2005 to 14.5 percent in 2007, and producing revenue of $5.974 billion for fiscal year 2007.

IBISWorld, "Changing Consumer Tastes Creates Explosive Growth for Domestic and International Bottled Water Brands—Revenue in 2007 Expected to Reach $5.974 Billion with Growth Set to Climb Higher Through 2012," May 21, 2008. Copyright © 2008, IBISWorld. Reproduced by permission.

Purified water is currently the leading global seller, with U.S. companies dominating the field. The U.S. is the largest consumer market for water [in] the world, followed by Mexico, China, and Brazil. Natural spring water, purified water, and flavoured water, have been identified by IBISWorld as the fastest growing segments.

"Premium" Water

"But there's more to the industry's strong performance than meets the eye, according to Senior Analyst with IBISWorld Mr. George Van Horn. "Because of the homogenous nature of the product, producers need to invest substantially in branding, advertising, and promotional activity to differentiate their offering, and to attract and retain consumers who would otherwise substitute readily between waters," said Mr. Van Horn. "As a result, we're seeing growth in the so-called 'premium' section of the industry, with some manufacturers promoting their water as superior in an attempt to extract higher margins." He added, "This has led to an increase in the market for imported products, as is demonstrated by the success of café-focused European brands such as Perrier and San Pellegrino, and the recent success of Fiji water in the U.S."

"In the U.S., the supply market is largely geared toward the production of purified water, to be sold in bulk—often through supermarkets and small retailing stores," said Mr. Van Horn. "And while this has spurred the dominance of the U.S. as a global market supplier, providing low cost, bulk purchase water to local and international markets, the growing trend towards fashionable, premium products may see more infiltration from imported brands within the U.S."

Mr. Van Horn explained that following trends abroad, bottled water had the potential to become as much a fashion accessory as a beverage, predicting savvy producers will establish niche operations supplying limited market segments with specialized and top-of-the line products.

Bottled water in the United States generates nearly $6 billion a year in revenue and is the fastest growing beverage market in the country.

"The current mediascape, particularly women's magazines, is saturated with images of celebrities flaunting premium water products in fashionable designer bottles," he said. "This has particularly been the case with Fiji water, with celebrity uptake no doubt contributing largely to the brand's success in the U.S."

"Females and younger consumers account for slightly larger levels of bottled water consumption, with media support behind a brand, the bottle design, and the label all playing a part," said Mr. Van Horn. "Women are also more diligent than men at drinking the recommended eight glasses of water a day, as well as being, on the whole, more health conscious."

"Functional" Water

Alongside premium waters, 'functional' water is another area that is driving industry revenue, with products making unique health claims targeting consumers who switch drinks during the day depending upon their immediate needs.

"The creation and promotion of sports waters and other near waters has helped bottled water win market share from high-sugar soft drinks, energy drinks, and sports drinks," . . . said Mr. Van Horn.

Functional waters—encompassing sports, flavoured, near and enhanced waters—compete as substitutes for soft drinks, as they are flavoured but do not have a high sugar content found in soft drinks. IBISWorld believes that as the industry matures and consumers become more informed, these sub-segments should become more clearly defined in the market.

"The rapid introduction of new products, and new packaging, make the bottled water market an extremely dynamic industry, and America's high level of carbonated soft drinks (CSD's), energy drinks, and sports drinks, and comparatively low-level of consumption of premium and enriched water products suggests that the U.S. market still has potential for a high rate of sales growth before reaching saturation," Mr. Van Horn added.

The bottled water industry has gained from the increasingly frantic pace of life. "With people trying to accomplish more each day, with less time for rest, and the rising preference for convenient snacks, dining out and takeaway meals, bottled waters are becoming an important convenient fact of life," explained Mr. Van Horn.

Some Brands Are Safer than Others

And though most brands of bottled water have been tested and shown to have no health benefits above those of tap water, many consumers won't be convinced. In fact, consumers are increasingly worried about the quality of bottled water that is often bottled using the same municipal water supplies that come out of home tap water.

"Following a recently released report based on a five-month undercover story, the Associated Press (AP) revealed that a vast array of pharmaceuticals have been found in the drinking water supplies of millions of Americans," said Mr. Van Horn. "What most consumers don't know is that most brands of bottled water undergo no filtration during the bottling process. Many bottled water brands are essentially the same quality that comes from everyday home tap water. Consumers are essentially buying a brand, a nice label, and the convenience."

Members of the AP Investigative Team reviewed hundreds of scientific reports, analysed Federal drinking water databases, visited environmental study sites, treatment plants, and interviewed more than 230 officials, academics and scientists across the U.S.

What they found was alarming. For example, officials in Philadelphia said testing there discovered 56 pharmaceuticals, or drug by-products, in treated drinking water, including medicines for pain, infection, high cholesterol, asthma, epilepsy, diabetes, heart disease, and mental illness. Sixty-three pharmaceuticals or by-products were found in the city's watersheds.

At a conference last summer [2007], Mary Buzby, director of environmental technology for Merck & Co. Inc., said: "There's no doubt about it, pharmaceuticals are being detected in the environment and there is genuine concern that these compounds, in the small concentrations that they're at, could be causing impacts to human health or to aquatic organisms."

"We know we are being exposed to other people's drugs through our drinking water, and that cannot be good," says Dr. David Carpenter, who directs the Institute for Health and the Environment of the State University of New York at Albany. While the bottled water industry is beginning to address this issue, according to the EPA [Environmental Protection Agency], currently [2008] there are no sewage treatment systems specifically engineered to remove pharmaceuticals.

According to Mr. Van Horn, not all bottled water is laced with various pharmaceutical drugs. "Brands like Fiji Water, Evian, and many other premium brands are bottled from water sources that are found in remote places such as in the mountains where pharmaceutical drugs from human waste don't exist."

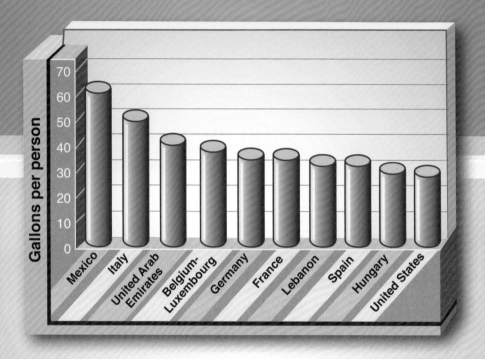

Taken from: John G. Rowdan Jr., "Challenging Circumstances Persist: Future Growth Anticipated," *Bottled Water Reporter*, p.10, April/May 2010.

Challenges Facing the Bottled Water Industry

Looking ahead beyond safety issues, Mr. Van Horn said there will be significant challenges facing the industry, mostly in the form of increasing infrastructure in developing countries, competition from premium labels and imports, and competition from low-calorie, sugar free carbonated soft-drinks (CSD's) and energy drinks, as well as dental care and environmental concerns, and mounting packaging costs.

Mr. Van Horn added, "Within the U.S. suppliers currently focus on the production of low cost purified water, a market that is largely supported by developing countries, and concerns about the quality of tap water rather than fashion or brand loyalty."

"In the coming years, due to developments in infrastructure in these countries, we may see great improvement in the quality of public water sources which may detract from further growth," predicted Mr. Van Horn.

"The U.S. is also still the major consuming country of soft drinks in the world with Americans consuming around 51.4 gallons of carbonated soft drinks per person each year, while at the same time Americans are becoming increasingly health-conscious. With that, the bottled water market increasingly finds itself competing with low-calorie, sugar free CSD's, and nutrient-enriched energy and so-called sports drinks.

"As consumption of bottled water by children increases, so too will concerns about the impact on their teeth—with tap water currently providing their main source of fluoride," said Mr. Van Horn. "Increasing environmental awareness, and concerns about the effects of manufacturing bottled water will also place pressure on the industry—with studies showing that it can take up to seven quarts of water and a quart of crude oil to produce about one quart of bottled water."

"In addition, over the next few years packaging costs, particularly for petroleum-based PET resin, will rise, putting pressure on profit margins. This is another reason why manufacturers will ramp up investment in developing higher priced premium spring waters and functional waters to partially offset cost pressures," Mr. Van Horn added.

Yet on the flip side, IBISWorld predicts these changing trends will surely see dominant U.S. players Coca-Cola Enterprises, Inc. and The Pepsi Bottling Group moving forward in the premium water market, creating new products focusing on nutrient enriched and flavoured water products and adopting fashionable packaging to appeal to the style-conscious. A move, which coupled with strong marketing campaigns, and the brand strength already achieved by these companies, could lead to U.S. premium water brands expanding more aggressively into the South American, Asian and Australian premium product markets.

Drinking Bottled Water May Hurt the Environment

Tom Paulson

> Writer Tom Paulson was a science reporter at the *Seattle Post-Intelligencer* for twenty-two years. He is the president of the Northwest Science Writers' Association. In this article Paulson examines the economic and environmental costs associated with the rapidly increasing consumption of bottled water. He notes a large amount of oil is required to make the billions of plastic bottles purchased each year. Additionally, he says many groups are concerned that bottled water is a step toward water privatization, which will drastically increase the cost of a basic resource.

America's infatuation with drinking high-priced "natural" water from a bottle rather than from the tap is contributing to global warming and could even qualify as an immoral act.

That, at least, is the position of a number of environmental, social justice and religious organizations.

Bottled Water Has High Energy Costs

"People need to think about all the unnecessary energy costs that go into making a bottle of water," said Peter Gleick, an expert on water policy and director of a think tank in Oakland, Calif., called the Pacific Institute.

Tom Paulson, "Thirst for Bottled Water May Hurt Environment," *Seattle Post-Intelligencer*, April 19, 2007. Copyright © 2007, *Seattle Post Intelligencer*. Reproduced by permission.

More than 8 billion gallons of bottled water is consumed annually in the U.S.—an 8-ounce glass per person per day—representing $11 billion in sales. The Earth Policy Institute estimated that to make the plastic for the bottles burns up something like 1.5 million barrels of oil, enough to power 100,000 cars for a year. Nearly 90 percent of the bottles are not recycled.

Gleick offered a simple way to visualize the average energy cost to make the plastic, process and fill the bottle, transport bottled water to market and then deal with the waste:

"It would be like filling up a quarter of every bottle with oil."

One of the simplest things folks can do to reduce their "energy footprint," he said, is to drink tap water rather than buy bottled water. If you don't like the taste, he said, buy a filter.

Bottled Water Is Not Healthier than Tap Water

"There's really no valid reason to think bottled water is any healthier than tap water," Gleick said. . . .

Despite the fact that the United States generally has high-quality tap water, it is the world's largest market for bottled water. There are a variety of explanations for this put forward by the purveyors of bottled water, including the contention that it is cleaner than tap water.

"It's about purity and convenience," said Trish May, chief executive officer of Athena Partners, a non-profit Seattle-based organization that produces Athena brand bottled water. "We're doubling our sales every year and now sell more than a million bottles a month."

Athena is one of the small, local bottled-water producers in the area. It is unique in this business—and perhaps more difficult to make a target of ecological outrage—because May, a breast cancer survivor, started selling bottled water to raise money for women's cancer research.

"We give every penny of our profits to cancer research," she said.

The water used by Athena—just as for Aquafina, Dasani and other brands—starts as plain tap water. It already has been through a purification process, but the water that will be put in bottles is further "purified" by a number of processes, such as filtration or reverse osmosis (which removes minerals that are then sometimes added back, mostly for taste reasons).

"I would submit to you that our purified water, with minerals added, is more pure than municipal water," May said.

That's not always going to be the case, said Gina Solomon, a senior scientist with the Natural Resources Defense Council.

"The bottled water industry is selling a vision of purity and people are buying it with the best of intentions," Solomon said.

"What they don't realize is that bottled water is actually much less regulated than tap water. There are a number of studies in which we find arsenic, disinfection byproducts and bacteria in bottled water."

The Food and Drug Administration, which is responsible for regulating bottled water, last month [March 2007] recalled Jermuk bottled water, sold in California under five brand names, after finding levels of arsenic high enough to cause nausea. But such recalls are unusual.

Differences in Regulation

The FDA does allow trace levels of contaminants in bottled water based on the same criteria set by the Environmental Protection Agency for tap water. But on the FDA's web site, the agency also says, "Bottled water plants generally are assigned a low priority for inspection."

The FDA is required to inspect water-bottling plants twice a year. In Washington, that duty is often delegated to inspectors with the state Department of Agriculture.

The FDA has a list of 44 firms it regulates here. State Agriculture officials listed 32 water "processors" they regulate and only about 20 of the firms are on both lists. That, officials said, may be because of the fact that some are just ice producers or that some have ceased operations.

"Also, if a firm does not engage in interstate commerce (receiving ingredients or shipping outside the state), it would not be con-

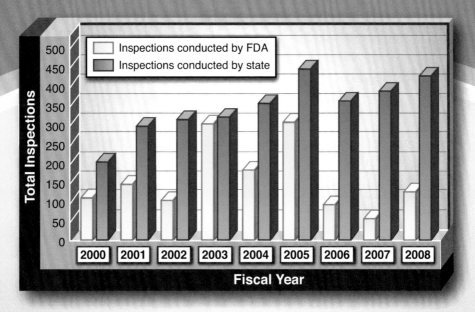

Bottled Water Facility Inspections Conducted by FDA and States

Taken from: U.S. Government Accountability Office analysis of Food and Drug Administration data.

sidered an FDA workload obligation," said Stephanie Dalgleish, with the Seattle office of the FDA. That means anyone making bottled water here and selling it only in-state is not regulated by the FDA.

There are about 4,000 municipal water systems in the state [Washington] that serve at least 25 people or more. These are regulated by the state Department of Health and the EPA on a near-constant basis.

"People are told within 24 hours if there's any problem, or potential problem, with their water system," said Leslie Gates of the health department's Office of Drinking Water.

A recent break-in at a water supply facility for the town of Orting, for example, prompted officials to suggest residents drink only bottled water until they could assure no contamination. There was none.

The water system for the City of Seattle, which also operates under EPA and Department of Health regulations, is monitored 24 hours a day, with constant sampling throughout the system and up in the wilds of the Cedar and Tolt watersheds.

"We never shut down," said Wylie Harper, water quality manager for the city. The water supplied to Seattle residents is purified through many of the techniques used for bottled water, so Harper joked that maybe the city should start bottling its water.

"But our focus isn't on making a profit," Harper said. "We provide a community service."

The bottled water market is big business. Coca-Cola (Dasani), PepsiCo (Aquafina) and Nestle (Perrier, Poland Spring and a host of other brands) are the major players in the United States.

Wall Street and investment managers are predicting the bottled water market (or, as one enthusiast called it, the "blue gold" market) will keep growing. Water, some financial investment managers say, is the next-best thing to oil or diamonds. And that's where the moral issues of bottled water come in.

Moral Objections to Water Privatization

The United Church of Christ, United Church of Canada, National Council of Churches, National Coalition of American Nuns and Presbyterians for Restoring Creation are among the religious organizations that have raised questions about the "privatization" of water.

They regard the industrial purchase and repackaging at a much higher resale price of this basic resource as an unethical trend. (Bottled water costs about 1,000 times more than tap water.)

"The moral call is for us to not privatize water," said Cassandra Carmichael, director of eco-justice programs for the National Council of Churches. Bottled water is the tip of the iceberg, Carmichael believes, in a push by industry to take ownership of this basic resource.

"We're scratching our heads on that one," said Preston Read, spokesman for the American Beverage Association. "Water priva-

Eight billon gallons of bottled water are sold in the United States each year. To manufacture the plastic bottles needed as containers requires 1.5 million barrels of oil. Nearly 90 percent of the bottles are not recycled.

tization is certainly a big issue but I don't see it as connected to bottled water."

As for the claim that bottled water causes global warming, Read said the same argument could be made against any beverage that is packaged in a plastic bottle, transported and sold.

"I think it's a little bit odd that bottled water is being singled out in this way," he said.

Ethos Water says its goal is to use profits to assist poor communities hard hit by the world water crisis. Ethos is a water bottler that was acquired in 2005 by Starbucks. Its founders say they launched the company a few years before that, in California, to raise money for water projects in the developing world.

Today, as a subsidiary of Starbucks, Ethos donates five cents for every bottle sold toward the goal of raising $10 million for water projects in poor countries.

"I wanted to create a brand that would raise awareness about the world water crisis," said Peter Thum, founder of Ethos Water and now a vice president at Starbucks.

Thum says he respects Gleick and understands his complaint about the energy costs that go into bottled water. He said he didn't know the economics of the situation well enough to respond to concerns about water privatization.

"I'm not going to defend the bottled water industry," Thum said. "Ethos Water can't answer for what others in the industry are doing. We're just trying to take the demand that is there and divert it to do some good."

Though not everyone accepts that Ethos Water is indeed focused more on doing good than making a profit, Ethos has already funded a number of water improvement projects in places such as Bangladesh, Indonesia, Ethiopia, Honduras and India.

But Athena and Ethos are hardly representative of the bottled water industry. Thum and May were willing to tackle these concerns, but most of the other bottlers and distributors contacted for this story did not respond.

Gleick said he is not opposed to water privatization, as long as the focus is on providing people with affordable access to water. But he and others are definitely opposed to the unnecessary use of bottled water because of its environmental impact.

But it is the demand for bottled water itself that many believe is bad.

"This is not an issue that's going to go away," Gleick said. "If anything, it's a growing movement. I think consumers deserve the option of drinking bottled water. But I also think they need to be informed about its true economic and environmental costs."

TWELVE

The Demand for Ethanol Is Creating a Water Problem

Jim Moscou

> Jim Moscou is a contributing writer for *Newsweek* magazine and an adjunct instructor at the University of Colorado at Boulder. Here, he explains how the rising demand for ethanol—an eco-friendly, or "green," energy source—has worsened the water shortage in eastern Colorado and Kansas. Farmers are growing more corn to produce ethanol, and that corn requires a lot of water. This has led to a political battle over water rights, Moscou says.

Mike Adamson remembers when water wasn't such a problem. As a kid growing up on his family's cattle feedlot along the Colorado-Kansas border, "you could dig a post hole and see water runnin' in the bottom," he recalls. Today [February 2008], Adamson is 48 and in charge of the family business, Adamson Brothers and Sons Feedlot, a holding ranch for cattle as they go to market. And the water, he says, is disappearing. "The lakes are gone. The wetlands are gone." In fact, Adamson adds, entire stretches of the nearby Republican River are gone.

In the arid regions of the American West, water has always been a precious, liquid gold. But in Adamson's home of Yuma County, two hours east of Denver, the stakes just got higher. Thanks to the boom in ethanol production spurred by green-

Jim Moscou, "Liquid Gold," *Newsweek*, February 21, 2008. Copyright © 2008, *Newsweek*. Reproduced by permission.

energy concerns, corn farmers in Yuma County—one of the top three corn-producing counties in the country—are enjoying a new prosperity.

Eco-friendly Fuel Comes with Costs

But the green-fuel boom touted as a clean, eco-friendly alternative to gasoline is proving to have its own dirty costs. Growing corn demands lots of water, and, in Eastern Colorado, this means intensive irrigation from an already stressed water table, the great Ogallala aquifer. One sign of trouble: in just the past two decades, farmers tapping into the local aquifers have helped to shorten the North Fork of the Republican River, which starts in Yuma County, by 10 miles. The ethanol boom will only hasten the drop further, say scientist and engineers studying the aquifers. The region's water shortage has pitted water-hungry farmers against one another. And lurking in the cornrows: lawsuits and interstate water squabbles could shut down Eastern Colorado's estimated $500 million annual ethanol bonanza with the swing of a judge's gavel. Collectively, "[ethanol] is clearly not sustainable," says Jerald Schnoor, a professor of engineering at the University of Iowa and cochairman of an October 2007 National Research Council study for Congress that was critical of ethanol. "Production will have serious impacts in water-stressed regions." And in Eastern Colorado, there's lots of water stress.

Still, with so much money growing in the fields, the current problems haven't stopped anyone on Colorado's plains. "Finally, here's the alternative market that farmers have been working toward for decades," said Mark Sponsler, executive director of the Colorado Corn Growers Association. The state's farmers planted a near record acreage of corn in 2007, up nearly 20 percent from the year before. It's not hard to see why. After hovering around $2 a bushel for nearly 50 years, corn is trading at about $4.50 today [February 2008]. Meanwhile, the [George W.] Bush administration has called for ethanol to displace 15 percent of the nation's gasoline supply by 2015, double that by 2030. And Yuma is preparing. The state's two ethanol plants have been built nearby in just the past few years, with a third on the way. "It sure is a good time," says Byron Weathers, a farmer with 2,500 acres of corn. "It's definitely been a big plus for our state. The whole nation, really."

Tensions over Water Rights

But the effort to keep the good times rolling locally has actually fueled a bitter Hatfield-vs.-McCoy atmosphere in these parts. "There's definitely tension between families," one long-time Yuma corn farmer said, who requested anonymity due to the sensitivity of the situation. Here's the trouble: eastern Colorado is painfully dry, but it sits on top of one of the world's largest underground freshwater oceans—the Ogallala Aquifer, which stretches

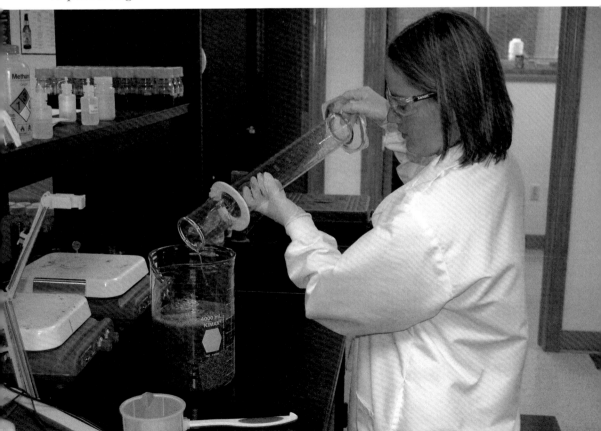

A research technician at an ethanol plant adds water to a corn mixture; the combined ingredients will produce ethanol. Ethanol production uses more water than the amount required to grow the corn.

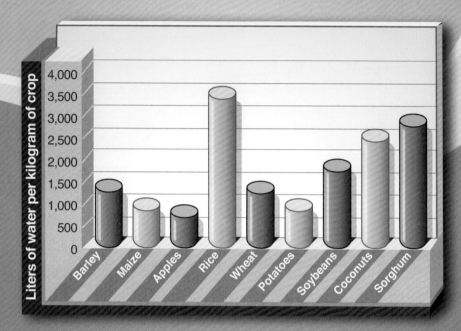

Taken from: www.waterfootprint.org.

from Montana to New Mexico. Seepage from the Ogallala in eastern Colorado creates the headwaters for the North Fork of the Republican River, which flows past the Adamson family farm, and into Nebraska and Kansas. But before the Republican reaches the border, 4,000 groundwater wells tap the Ogallala, which depletes the river further and faster than rain or winter run-off can recharge it. Near Yuma, the water table has dropped more than 100 feet in the past few decades, drying out Adamson's post holes.

In Yuma County, the battle is between farmers who irrigate 400,000 corn acres with groundwater against those who draw surface water from the river using drainage ditches, like Adamson. (Adamson uses the water to grow less-water intensive crops, like wheat, that he can feed to the cattle.) As the wells draw down the water table, the river flow drops, too. So, when the valves are opened, the water barely trickles into irrigation ditches, like

Adamson's, whose family's right to draw that water according to state law dates back to the 1800s. "We're the canary in the coal mine," Adamson said. If there's little water in his ditches, the river is running low.

To be sure, scientists have been watching the depletion of the Ogallala for decades. Years of drought haven't helped either. But the corn-based ethanol boom has added pressure, and money, to keep the tap on. So to save the river and their water, Adamson and a group of surface water-right holders sued in 2005 to shut off the wells. A hearing is set for June. If they win, hundreds, maybe thousands of groundwater wells irrigating corn could be shut off instantly. "It would devastate the economy," says Doug Sanderson, the city manager of Yuma, the county seat.

Yuma County farmers face another water threat, this one from neighboring Kansas. The downstream state has struggled for decades to get its fair share of the Republican's waters. Tensions peaked eight years ago when Kansas brought a lawsuit against Colorado and Nebraska to the U.S. Supreme Court—and won. Today, the two states still owe Kansas enough water to supply a small city for a year. But, like a shop-a-holic with credit cards, Colorado's groundwater wells keep pumping. "We're at a junction with the interstate compact," says Dave Barfield, chief engineer for Kansas. "[Kansas] farmers are being hurt. They are telling me to go get 'em. . . . And we are." Last month, Kansas demanded its water, suggesting Colorado and Nebraska shut down groundwater wells. If things get worse, the Supreme Court could order it. The threat has sent Colorado's politicians, farmers and others scrambling, and proposed solutions are as perplexing as the problems.

No Easy Solutions

To send Kansas its water—and keep the Colorado well on—a state legislator is pushing to drain the Bonny Reservoir, a popular border lake called the "crown jewel" of eastern Colorado. It's a key stopping point for migratory birds, a fishery maintained by the state, and leased by Colorado from the federal government, who are not likely to let the water go. Still, the bill's sponsor, state

Sen. Greg Brophy of Yuma, has made the message clear: "We can't value fish over farmers."

Yuma corn farmers have come up with their own idea. Last month, the local Republican River Water Conservation District, a board responsible for keeping Colorado in water compliance with Kansas, approved the funding for a multi-million dollar pipeline that will pump water into the Republican River from a farm willing to retire 6,000 acres. Water will flow to Kansas. Problem solved. The source of that water? The Ogallala Aquifer. It's an idea some have called robbing Peter to pay Paul. "It is to a degree," says Ken Knox, Colorado's chief deputy state engineer. "But we're trying to maintain the entire social-economic production in this part of Colorado."

What's becoming clear is that the price to keep ethanol profitable is not cheap. The purchase of those wells will cost more than $50 million—a market-maker price tag that's even catching the eye of the surface-water right owners. "You know, money is an enticing thing," Adamson said. "It's great to be noble. Sometimes it's hard to be noble. But you've got to take care of your family." One attorney close to the case is more succinct: "[Surface-water owners] are probably just waiting for the right price." Should the right price come along, Ogallala's groundwater will be left uncontested, at least in Colorado, a likely scenario. As for the Republican River? "We know we have a finite resource. We know it won't last forever," says Yuma City Manager Sanderson. "But we certainly don't respect the resource more than we respect the people."

Scientists and engineers say there's a clear lesson from the Republican River saga: water and energy are inextricably linked. "They will be the two driving forces of the future," says Knox. "And we're starting to see the future in this region." Professor Schnoor calls ethanol simply "a bridge fuel" to undiscovered and truly environmentally-friendly technology. Meanwhile . . . Adamson is frustrated. "Trying to solve problems by using the same old techniques doesn't solve the problem," Adamson says. "We're going to make the area a desert. It's going to be uninhabitable." And that would be a high price to pay.

APPENDIX

What You Should Know About Water

Water Scarcity
- A 2010 report by the World Health Organization (WHO) determined that 2.6 billion people do not use improved sanitation.
- The WHO report concluded that 884 million people do not use improved sources of drinking water.
- The Intergovernmental Panel on Climate Change (IPCC) predicts that by the year 2050 around 60 percent of the world's population will experience severe water shortages.
- About 33 percent of the world's population is already under water stress, according to the IPCC.

Laws and Regulations
The United States has enacted numerous laws to address water pollution:

- The Clean Water Act bans the discharge of pollutants into the water and specifies "pollutants" to include, among other things, solid waste, sewage, garbage, chemical waste, biological waste, radioactive materials, rock, sand, and agricultural waste. The Clean Water Act does *not* ban ships from dumping human waste, nor does it ban US Navy ships from discharging munitions.
- The Ocean Dumping Act recognizes the harm of dumping and bans the *transport* of waste for the purpose of dumping it into the ocean.

- The Oil Pollution Prevention, Response, Liability, and Compensation Act requires ships that spill oil to pay for the damage and cleanup costs. The law also requires oil tankers to have double hulls, which reduce the risk of spills.
- The Coastal Zone Management Act recognizes that coastal zones are vulnerable and requires coastal states to manage pollution.

Bottled Water
- Water has been collected and bottled for centuries.
- Americans consume 8.6 billion gallons of bottled water each year.
- The cost of bottled water is up to ten thousand times the cost of tap water.
- Forty percent of all bottled water is taken from municipal water sources (tap water).
- It takes three times as much water to produce a plastic bottle as it does to fill it.
- Seventeen million barrels of oil are used each year to produce water bottles.
- Only one in five water bottles is recycled.

The Earth's Water Supply
- Only 3 percent of earth's water is freshwater, while 97 percent of the water on earth is salt water.
- The water found at the earth's surface in lakes, rivers, streams, ponds, and swamps makes up only 0.3 percent of the world's freshwater.
- A total of 68.7 percent of the freshwater on earth is trapped in glaciers.
- Thirty percent of freshwater is in the ground.
- There is more freshwater in the atmosphere than in all of the rivers on the planet combined.
- If all of the water vapor in the earth's atmosphere fell at once, distributed evenly, it would only cover the planet with about an inch of water.

Water Usage
- American residents use about 100 gallons of water per day.
- European residents use about 50 gallons of water per day, according to the World Water Council (WWC).
- The WWC also reports that residents of sub-Saharan Africa use only between 2 and 5 gallons of water per day.
- Americans use more water each day by flushing the toilet than they do by showering or any other activity.
- The average faucet flows at a rate of 2 gallons per minute.
- *The New York Times* reports that the New York City water supply system leaks 36 million gallons per day.
- It takes more water to manufacture a new car (39,090 gallons) than to fill an aboveground swimming pool.

What You Should Do About Water

Because the water crisis is a global problem, it may seem like one person cannot make a difference. It is true that the issue needs to be dealt with on a large scale: Governments can make laws and regulations, big corporations can change their policies, and new technologies can fix distribution and hygiene problems. However, we all can do our part. If everyone does a little it will add up to a lot. Here are some things you can do to help conserve water, reduce pollution, and raise awareness.

Conserving Water at Home

A good way to start saving water is to use less on a daily basis. Saving a gallon here and a gallon there can add up, especially if everyone in your household contributes. Many of these steps are easy to do; they just take a little awareness. For instance, turning off the water while brushing your teeth can save 25 gallons per month. Combine this with turning off the water while shaving—300 gallons per month—and while washing your hair—150 gallons per month—and you can greatly reduce your family's water consumption, especially if you convince your family members to join in. Perhaps the easiest way to conserve water in the bathroom is to shorten your showers. A minute or two less each time can save 150 gallons per month. This adds up to a substantial savings, and that is just one room in your home.

You can also save water during many household chores. Use a broom instead of a hose to clean your driveway. When washing dishes by hand, do not leave the water running: fill one sink with wash water and the other with rinse water. If your house has a dishwasher, only run it when it is full. The same goes for your washing machine. Using these appliances conservatively can save up to 1,000 gallons of water every month. Using a hose nozzle or turning the water off when you wash your car will save up to 100 gallons every time. Do not water your lawn on windy days, when most of the water will blow away or evaporate.

Finally, your house may be wasting water even when nobody is using the hose or faucets. The slow dripping of leaky pipes can add up. Grab a wrench and fix the leaks (or ask a parent to do so) to save up to 140 gallons per month.

Another way to conserve water is to be mindful of the products you consume. The Water Footprint Network (www.waterfootprint.org) has researched how much water goes into the production of various foods and personal goods. Try to avoid products that require large amounts of water to create. For example, try eating less beef; a typical hamburger takes 630 gallons of water to produce.

Altering your daily habits can conserve a lot of water, but households can create additional savings through more permanent measures. You could discuss the possibility of making a major change. For instance, low-volume or dual-flush toilets use considerably less water than do traditional toilets. Landscaping is another potential water waster. Choose plants appropriate for your climate; they can thrive in the natural environment and will need less watering. If your house has a pool, use a pool cover. The cover will reduce water loss from evaporation.

Additional ways you can save water are available at the website of the organization Water Use It Wisely (www.wateruseitwisely.com).

Preventing Pollution

While a lot of pollution comes from major industries, families can still do their parts to keep pollution out of their local water systems. Be aware of what you are putting down the drain or into the sewers.

Keep litter, pet wastes, leaves, and debris out of street gutters and storm drains—these outlets drain directly to lakes, streams, rivers and wetlands. Instead, use the yard waste to start a compost pile. Do not pour household goods such as cleansers, beauty products, medicine, paint, and lawn care products down the drain. If you have a car, be sure to clean up spilled brake fluid, oil, grease, and antifreeze. Do not hose them into the street where they will eventually reach local streams and lakes.

Compared to the big industrial polluters, these steps may not seem like much. However, if the millions of households in America take these actions, the changes become significant. Additionally, much of your household pollution will wind up in your local water source; reducing pollution helps keep *your* water supply clean!

Bottled Water

Bottled water is portable and convenient but can be very harmful to the environment. A lot of energy is used to create and transport the bottles, many of which end up in landfills. Be sure to recycle empty water bottles. Additionally, some companies now use biodegradable plastic in their bottles. Purchase these brands to support this environmentally friendly practice.

A major benefit of bottled water is the convenience of portability. Drink tap water at home or in restaurants, and use bottled water only when you are on the go. Better yet, consider investing in a reusable bottle, and fill it from the tap before you go out. It will quickly pay for itself; every time you refill you are saving the cost of a bottle of water.

Bottled water is not always cleaner and safer than tap water. Before purchasing a brand of bottled water, research where the water comes from and what testing and filtration procedures the company uses. You can also find out how clean your local tap water is. Public water systems are required to release testing data. Contact your local government or the Environmental Protection Agency for information.

Get Involved

Whereas drinking bottled water is an individual decision, to make a difference on a larger scale, there are many organizations devoted to fighting water pollution and water shortages. Contact one in your area for information on what you can do to help. Many of them have volunteer opportunities for students to assist with cleanup or campaign efforts.

You can also become a part of a group to make your voice heard. If your school has an environmental club that deals with water issues, join it. If not, work with some friends to start one. As a club, you can write to your city council members or congressperson to express your concerns about water-related issues and work with these representatives to address these problems.

Finally, consider whether a career relating to water conservation and hygiene interests you. As a legislator, a scientist, an activist, or one of many other professions, you can work every day to fix the world's water problems.

ORGANIZATIONS TO CONTACT

Alliance for Water Efficiency
300 W. Adams St., Ste. 601
Chicago, IL 60606
phone: (866) 730-2493
fax: (773) 345-3636
e-mail: jeffrey@a4we.org
website: www.allianceforwaterefficiency.org

The Alliance for Water Efficiency is a nonprofit organization dedicated to the efficient and sustainable use of water. Located in Chicago, the Alliance serves as a North American advocate for water efficient products and programs and provides information and assistance on water conservation efforts.

Charity: Water
200 Varick St., Ste. 201
New York, NY 10014
phone: (646) 688-2323
fax: (646) 638-2083
e-mail: info@charitywater.org
website: www.charitywater.org

Charity: Water is a nonprofit organization bringing clean and safe drinking water to people in developing nations. Founded in 2006, it has funded thousands of projects in seventeen countries. Charity: Water uses both mainstream and social media platforms to raise awareness.

A Child's Right
1127 Broadway, Ste. 102
Tacoma, WA 98402
phone: (253) 327-1707
e-mail: acr@achildsright.org
website: www.achildsright.org

A Child's Right is a nonprofit relief organization focused on helping impoverished children around the world. Its members come from very diverse backgrounds and work with water purification and sanitation technologies, water quality, hygiene education practices, pediatric care, and international relief work. Its goal is to use technology and education to provide children with clean water and hygiene education.

Clean Water Action
1010 Vermont Ave. NW, Ste. 400
Washington, DC 20005-4918
phone: (202) 895-0420
fax: (202) 895-0438
e-mail: cwa@cleanwater.org
website: www.cleanwateraction.org

Founded in 1972, Clean Water Action is an organization of 1.2 million members working to empower people to take action to protect America's waters and build healthy communities. The organization has succeeded in winning some of the nation's most important environmental protections through grassroots organizing, expert policy research, and political advocacy focused on holding elected officials accountable to the public.

Freshwater Society
2500 Shadywood Rd.
Excelsior, MN 55331
phone: (952) 471-9773
fax: (952) 471-7685
e-mail: freshwater@freshwater.org
website: www.feshwater.org

Founded in 1968, the Freshwater Society is a Minnesota-based nonprofit organization dedicated to educating and inspiring people to value, conserve, and protect all freshwater resources. The society disseminates information through its newsletter and blog and sponsors forums, exhibits, and scholarships to raise awareness of freshwater issues.

National Resources Defense Council (NRDC)
40 West 20th St.
New York, NY 10011
phone: (212) 727-2700
fax: (212) 727-1773
e-mail: nrdcinfo@nrdc.org
website: www.nrdc.org

The NRDC is an environmental action organization that uses law, science, and the support of 1.3 million members and online activists to protect the planet's wildlife and wild places and to ensure a safe and healthy environment for all living things. With the support of its members and online activists, the NRDC works to solve the most pressing environmental issues we face today: curbing global warming, getting toxic chemicals out of the environment, moving America beyond oil, reviving our oceans, saving wildlife and wild places, and helping China go green.

Pacific Institute
654 Thirteenth St.
Preservation Park
Oakland, CA 94612
phone: (510) 251-1600
fax: (510) 251-2203
e-mail: info@pacinst.org
website: www.pacinst.org

The Pacific Institute works to create a healthier planet and sustainable communities. Its aim is to find real-world solutions to problems like water shortages, habitat destruction, global warming, and environmental injustice. Based in Oakland, California, the Pacific Institute conducts research, publishes reports, recommends solutions, and works with decision makers, advocacy groups, and the public to change policy.

US Environmental Protection Agency (EPA)
Ariel Rios Building
1200 Pennsylvania Ave. NW

Washington, DC 20460
phone: (202) 272-0167
website: www.epa.gov

The EPA is an agency of the US federal government created to protect human health and the environment. The EPA works with Congress to implement environmental laws and regulations. The agency conducts environmental assessment, research, and education. It employs seventeen thousand people among its national headquarters, ten regional offices, and twenty-seven laboratories across the country. More than half of the staff is engineers, scientists, and environmental protection specialists; the remainder includes legal, public affairs, financial, and computer specialists.

US Geological Survey—Water Resources Division (WRD)
USGS National Center
12201 Sunrise Valley Dr.
Reston, VA 20192
phone: (888) 275-8747
e-mail: contact via online feedback form at http://water.usgs.gov/user_feedback_form.html
website: water.usgs.gov

The WRD is a branch of the US Geological Survey. Its mission is to collect and disseminate reliable, impartial, and timely information necessary to understand the nation's water resources. The WRD is a workforce of thirty-four hundred people located in all states and territories at 181 offices, working with about fifteen hundred state and local agency cooperators. Information on local offices is available at the WRD website.

Water Footprint Network (WFN)
c/o University of Twente
Horst Building
PO Box 217
7500 AE Enschede
The Netherlands
phone: +31 53 489 4320

fax: +31 53 489 5377
e-mail: info@waterfootprint.org
website: www.waterfootprint.org

The WFN is a nonprofit foundation under Dutch law. It is an international network by and for its partners. The mission of the WFN is to promote the transition toward sustainable, fair, and efficient use of freshwater resources worldwide by advancing the concept of the "water footprint," a simple indicator of direct and indirect water use of consumers and producers.

World Health Organization (WHO)
Avenue Appia 20
1211 Geneva 27
Switzerland
phone: +41 22 791 21 11
fax: +41 22 791 31 11
e-mail: info@who.int
website: www.who.int

The WHO is the directing and coordinating authority for health within the United Nations system. It is responsible for providing leadership on global health matters, shaping the health research agenda, setting norms and standards, articulating evidence-based policy options, providing technical support to countries, and monitoring and assessing health trends.

BIBLIOGRAPHY

Books

Lono Kahuna Kupua A'o, *Don't Drink the Water (Without Reading This Book): The Essential Guide to Our Contaminated Drinking Water and What You Can Do About It*. Dorset, UK: Lotus, 2004.

Maude Barlow, *Blue Covenant: The Global Water Crisis and the Coming Battle for the Right to Water*. New York: New Press, 2007.

Maude Barlow, *Blue Gold: The Battle Against Corporate Theft of the World's Water*. Abingdon, Oxford, UK: Earthscan, 2003.

Cynthia Barnett, *Blue Revolution: Unmaking America's Water Crisis*. Boston: Beacon, 2011.

Francis H. Chapelle, *Wellsprings: A Natural History of Bottled Spring Waters*. Piscataway, NJ: Rutgers University Press, 2005.

Brian Fagan, *Elixir: A History of Water and Humankind*. London: Bloomsbury, 2011.

Peter H. Gleick, *Bottled and Sold: The Story Behind Our Obsession with Bottled Water*. Washington, DC: Island, 2010.

Robert Glennon, *Unquenchable: America's Water Crisis and What to Do About It*. Washington, DC: Island, 2010.

Stephen Hoffman, *Planet Water: Investing in the World's Most Valuable Resource*. Hoboken, NJ: Wiley, 2009.

Norris Hundley Jr., *The Great Thirst: Californians and Water—a History*. Berkeley: University of California Press, 2001.

Susan J. Marks, *Aqua Shock: The Water Crisis in America*. New York: Bloomberg, 2009.

Ken Midkiff, *Not a Drop to Drink: America's Water Crisis (and What You Can Do)*. Novato, CA: New World Library, 2007.

Fred Pearce, *When the Rivers Run Dry: Water—the Defining Crisis of the Twenty-First Century*. Boston: Beacon, 2007.

Peter Rogers and Susan Leal, *Running Out of Water: The Looming Crisis and Solutions to Conserve Our Most Precious Resource*. New York: Palgrave Macmillan, 2010.

Elizabeth Royte, *Bottlemania: How Water Went on Sale and Why We Bought It*. New York: Bloomsbury USA, 2008.

Steven Solomon, *Water: The Epic Struggle for Wealth, Power, and Civilization*. New York: Harper Perennial, 2010.

Kenneth M. Vigil, *Clean Water: An Introduction to Water Quality and Pollution Control*. Corvallis: Oregon State University Press, 2003.

Periodicals and Internet Sources

Randal C. Archibold, "A Century Later, Los Angeles Atones for Water Sins," *New York Times*, January 1, 2007. www.nytimes.com/2007/01/01/us/01water.html.

Jennifer Barone, "From Toilet to Tap," *Discover*, May 2008. http://discovermagazine.com/2008/may/23-from-toilet-to-tap.

E.A. Barrera, "Overcoming the Stigma of 'Toilet-to-Tap' Water," *San Diego News Network*, May 27, 2009. www.sdnn.com/2009-05-27/special-sections/water/overcoming-the-stigma-of-toilet-to-tap-water.

Jeff Conant, "Lords of Water." *E—the Environmental Magazine*, February 28, 2010. www.emagazine.com/archive/5067.

Erin Cunningham, "World Water Day: Thirsty Gaza Residents Battle Salt, Sewage," *Christian Science Monitor*, March 22, 2010.

John Kerry, "Why It's Clear There Is a Worldwide Water Crisis," Change.org, October 15, 2010. http://news.change.org/stories/why-its-clear-there-is-a-worldwide-water-crisis.

Maggie Koerth-Baker, "Is There Really a Water Crisis?," *Boing Boing*, November 16, 2009. www.boingboing.net/2009/11/16/is-there-really-a-wa.html.

Kathryn Kranhold, "Water, Water, Everywhere . . . ," *Wall Street Journal*, January 17, 2008. http://online.wsj.com/article/SB120053698876396483.html?mod=googlenews_wsj.

David Krantz and Brad Kifferstein, "Water Pollution and Society," University of Michigan. www.umich.edu/~gs265/society/waterpollution.htm.

Mayo Clinic Staff, "Water: How Much Should You Drink Every Day?," Mayo Clinic. www.mayoclinic.com/health/water/NU00283.

National Geographic, "Water: A Special Issue," April 2010. http://ngm.nationalgeographic.com/2010/04/table-of-contents.

Heather Sharp, "Gaza Thirsts as Sewage Crisis Mounts," BBC News, October 27, 2009. http://news.bbc.co.uk/2/hi/middle_east/8327146.stm.

Cristen Tilley, "Water Supply Crisis a Myth, Says Provider," Australian Broadcasting Company, May 17, 2006. www.abc.net.au/water/stories/s1640772.htm.

Bryan Walsh, "Earth Day: Are We Destroying the Oceans?," *Time*, April 14, 2010. www.time.com/time/health/article/0,8599,1982015,00.html.

Kenneth R. Weiss and Usha Lee McFarling, "Altered Oceans," *Los Angeles Times*, July 30–August 3, 2006. www.latimes.com/news/local/la-oceans-series,0,7783938.special.

Chaoquing Yu, "China's Water Crisis Needs More than Words," *Nature*, February 16, 2011. www.nature.com/news/2011/110216/full/470307a.html.

INDEX

A
Abu Taha, Ali, 31
Abu Taha, Mohammed, 31–32
Adamson, Mike, 79, 82–83, 84
ADC (Affordable Desalination Coalition), 50
Affordable Desalination Coalition (ADC), 50
Africa, 39–46
Al-Agha, Mohammed, 36, 38
Agriculture
  as largest consumer of freshwater, 16–17
  loss of water due to irrigation, 21–22
  water pollution from, 18–19
Alois, Paul, 15
Aquifers, pollution of, 18
Aristotle, 49–50
Arlosoroff, Saul, 38
Ashour, Ehab, 35
Athena Partners, 73, 78

B
Barfield, Dave, 83
Barnett, Tim, 5
Beckman, James, 51, 52
Besor Reservoir (Israel), 64
Binayo, Aylito, 39–40, 42, 46
Bonny Reservoir, 83–84
Borale, Birhane, 41
Bottled water
  annual per person consumption of, 70
  annual US consumption of, 73
  as fashion trend, 66–67
  inspection of facilities by FDA/states, 75
  may hurt environment, 72–78
  size of market, 65
Bottled water industry
  major players in, 76
  size of, 65–71
Bresnahan, Mike, 8
Brooks, David, 36
Brophy, Greg, 84
Buzby, Mary, 69

C
California, 5–9
Carmichael, Cassandra, 76
Carpenter, David, 69
Chapman, Michelle, 48, 52, 53
Clean Water Act (1972), 11, 13
Colorado River, 5–6
Community Led Total Sanitation initiatives, 58–59
Conservation, 8–9
  setting price on water, 30
Cook, Bob, 8
Crops, 82

D
Dalgleish, Stephanie, 75
Desalination, 21
  concerns with, 7–8, 62
  global capacity of, 52

scientific advances in, 47–53
in Western US, 6–8
Dewvaporation, 51
Dickinson, Elizabeth, 28
Drinking water
  access to, 23–27, 28–30
  numbers using unimproved sources of, 26
  pharmaceuticals in, 62–63, 69
  recycling sewage water, 60–64
Drought, 42

E
Earth Policy Institute, 73
Elimelech, Menachem, 51
End Water Poverty coalition, 59
Environmental Protection Agency (EPA), 63, 69
Estiphanos, Israel, 40–41
Ethanol, 79–84
Ethiopia, 39–46
Ethos Water, 78

F
Faeth, Paul, 43, 46
FDA. *See* Food and Drug Administration, US
Food and Drug Administration, US (FDA)
  inspection of bottled water facilities by, 75
  regulation of bottled water by, 74–75
Food and Water Watch, 23
Forward osmosis, 51
Freshwater
  agricultural consumption of, 16–17
  global distribution of, 37
  industrial consumption of, 17–18
  international conflicts over, 20–21
  lack of, 40–42
  number of people lacking adequate supply of, 15, 40
  residential consumption of, 18
  uses of, 16
Frost, Barbara, 54
Functional waters, 68

G
Gates, Leslie, 75
Gaza
  desalination and, 36–37
  international help for, 38
  sewage problems in, 34–35
  water shortage in, 31–33
Gleick, Peter, 72, 73, 78
Global warming, 5–6
Green Revolution, 17
Groundwater Replenishment System, 60–61

H
Hadera desalination plant (Israel), 49
Hagos, Tsega, 41–42
Harper, Wylie, 76
Health problems, 54–59
Hinkenein, Thomas, 48, 49
Hippocrates, 49–50
Howitt, Richard, 6
Human rights, access to water
  as, 23–27
  will not improve access, 28–30

I

IBISWorld, 65
International Food Policy Research Institute (IFPRI), 19
Irrigation, 16–17
  drip, 38
  greater efficiency in, 8–9
  loss of water through, 21–22

J

Jalto, Guyo, 45
Joseph, Bundaa, 55
Juma, Rideana, 57

K

Knox, Ken, 84

L

Lake Cumo, *11*
Lake Mead, 5
Lake Powell, 6
Lemeta, Wako, 45
Los Angeles
  opposition to sewage recycling in, 61
  water rationing in, 8

M

MacHarg, John, 50, 51
May, Trish, 73, 78
Mayes, Anne, 49
McKee, Mac, 35
MDG (Millennium Development Goal), 55–57
Membrane distillation, 51–52
Middle East
  conflicts over water in, 20–22
  suffers from water crisis, 31–38
Millennium Development Goal (MDG), 55–57
Moscou, Jim, 79
Mother Jones (magazine), 10

N

National Research Council, US, 48
Nile River, 20–21

O

Obama, Barack, 27
Oceans, 10–14
Ogallala Aquifer (US), 61, 80, 81–82, 84

P

Pacific Institute, 8, 72
Pallant, Eric, 33
Paulson, Tom, 72
Pharmaceuticals, 62–63, 69
Pollution
  from agricultural/industrial sources, 18–19
  from human waste, 54–55
  oceans are endangered by many sources of, 10–14
  from pharmaceuticals, 62–63, 69
Population growth, 16, 22, 47
Privatization, 22
  objections to, 76–78

R

Recycling, 60–64
Reverse osmosis (RV) desalination, 48, 50, 52–53
Rosenberg, Tina, 39

RV (reverse osmosis) desalination, 48, 50, 52–53

S
Sadat, Anwar, 20
Safieh, Abu, 36, 38
San Diego, 61
San Joaquin Delta (CA), 6, 7
Sanders, Jerry, 61
Sanderson, Doug, 83, 84
Sanitation, 54
Schnoor, Jerald, 80
Schwarzenegger, Arnold, 27
Seawater desalination. *See* Desalination
Service, Robert F., 47
Sewage/sewage systems
  In Gaza, 31–33
  health problems and, 54–59
  improved, 58
  recycling water from, 60–64
Sharon, Ariel, 20
Sirkar, Kamalesh, 51–52
Soft drinks, 71
Solomon, Gina, 74
Sponsler, Mark, 80

T
Thum, Peter, 78

U
UNEP (United Nations Environmental Program), 18
UNICEF (United Nations Children Fund), 24–25
United Nations Environmental Program (UNEP), 18

V
Van Horn, George, 66, 67, 68, 69, 70–71

W
Water
  global average use of, for crops, *82*
  home uses of, 63
  household members collecting, *41*
  use of, *17*
  See also Bottled water; Drinking water; Freshwater
Water rationing, 8
Water shortages
  can lead to food scarcity, 19
  is a global crisis, 15–22
  in Western US, 5–9
Water Stress Index, 16
Water-use efficiency, 8–9
WaterAid, 43
Weathers, Byron, 81
WHO (World Health Organization), 15, 24–25
World Bank, 15
World Health Organization (WHO), 15, 24–25
World Water Forum, 25

Y
Al-Yaqoubi, Ahmad, 34, 35, 37

Z
Zimmerman, Eileen, 6

PICTURE CREDITS

AP Images/Gary Kazanjian, 20
AP Images/Dirk Lammers, 81
AP Images/Emilio Morenatti, 32
Ron T. Ennis/MCT/Landov, 11
Gale, Cengage Learning, 17, 26, 37, 41, 52, 58, 63, 70, 75, 82
Sarah Beth Glicksteen/The Christian Science Monitor/Getty Images, 67
Val Handumon/EPA/Landov, 24
© Joe Hula/Alamy, 77
Jonathan Nackstrand/AFP/Getty Images, 64
Reuters/Landov, 7
Kfir Sivan/Israel Sun/Landov, 49
© Jim West/Alamy, 29
Xinhua/Landov, 44, 56